Remote Manipulation Systems

Quality Evaluation and Improvement

by

LEONID I. SLUTSKI
Ben-Gurion University of the Negev,
Beer-Sheva, Israel

SPRINGER SCIENCE+BUSINESS MEDIA, B.V.

A C.I.P. Catalogue record for this book is available from the Library of Congress.

ISBN 978-94-010-6166-7 ISBN 978-94-011-5153-5 (eBook)
DOI 10.1007/978-94-011-5153-5

Printed on acid-free paper

To Larissa and children Marina and Denis

TABLE OF CONTENTS

FOREWORD

A famous French writer, Anatole France, liked to say, "The future is a convenient place to position our dreams" (1927). Indeed, this remark gains full meaning when one considers the history of what we call today "Robotics."

For more than 3000 years, mankind has dreamt of the possibility of artificial machines that would have all the advantages of human slaves without any of their drawbacks.

With the developments in technology since the end of World War II, mainly with the explosive progress of computers, it was thought we might at last succeed in transforming this everlasting dream into reality. In the mind of scientists of the 1950's, to make such intelligent and autonomous machines before the year 2000 seemed a small challenge: it was obvious, thanks to computers and Artificial Intelligence.

But, in spite of progress in some directions, we must admit that the dream remains a dream and that the basic problems denying us a successful issue are not solved. In fact, if we except industrial robots, only calling for classical automata theory, the main advanced result concerning autonomous and intelligent machines is related to some understanding of reasons why we have failed during the past years.

One reason for this failure probably rests with thinking (but in refusing to admit it) of a robot mainly in terms of an artificial man, using, at a decisive level, properties of his mind or brain rather than properties of his body. Indeed, a robot is devoted to the material transformation of its environment; it is devoted to work. And work calls more for dexterity and training than for intelligence.

Similarly, to rely on a human model drives us to focus on that specific property of man which is his capacity not to have to take into account in advance the nature and properties of the environment or of the task he has to perform. In one glance and on-line, he is able to understand the whole problem he has to solve in order to execute the task, and he is able to instantaneously adapt his behaviour toward achieving his goal. As a consequence, the robot was designed as an isolated machine that did not take advantage of its interaction with the environment to perform the desired task.

As a recent example of the errors, we can cite the belief in computer simulation and calculation, first, as an exact representation of physical reality, and, second, as a convenient method to correcting any defect of a mechanical system under computer control.

So, summarizing the past history of Robotics, we find on one hand, this basic, unsuccessful search to give autonomy to robots, and on the other hand, people who need this autonomy for their applications but are obliged to let man into the control loop while they wait for the successful outcome; and, finally, industrial robots that are to some extent classical machines, from a control viewpoint.

In fact, these three main ways hide the fact that the man-concept representation and man-machine relations are the starting points of any progress in the field.

But, if today a "human centered system" appears as a potentially fructuous idea which mobilizes a great part of robotics research activity, that viewpoint was not obvious during the period spreading from the sixties to the eighties. Indeed, industrial engineers and managers believed in full automation of production systems, and the zero-man factory was a goal based on a huge hope of increasing profitability and a trust in the capacity of Robotics and related topics to reach the goal within a short period. But two big obstacles arose. The first one was the high cost of full automation to ensure reliability, safety, and maintenance with machines that have a weak adaptivity. Technologically speaking, we were not ready to win the challenge. The second obstacle was sociological. In fact, to drastically reduce the number of workers generates new problems. On the one hand, it would be more expensive than the previous situation when the cumulative effect of the financial burdens of unemployed people is estimated over a given whole country. On the other hand, exclusion of a lot of people from the job market demands moral responsibility and difficult policy problems.

Consequently, by the end of the eighties, a reaction to the former direction of research began to demand that a different approach to the problems of Robotics, including the presence of man into the process, be considered. But the question is: What are the best ways to go forward?

To exhibit the goal of this book, I am going to give a short panorama of Robotics research today. A relatively new idea consists in not starting with the objective to make an universal robot able to successfully achieve any task, but to start with the desired task.

A first case takes place when the task is imposed and must be performed by one or more robots. We have then three main directions possible: first, the task is relevant to industrial problematics and we can use conventional industrial robots that are only kinds of automata; second, the task is more complex and we need an autonomous robot we cannot make today; or, lastly, we choose to perform the desired task through a teleoperated system. In this last case, we shall use, in part, solutions presented in this book, either included into a conventional approach, or included in the general direction of mastering man-machine interactions, relying on these new techniques called Virtual Reality.

Whatever the status of the desired task (imposed or not imposed), making an autonomous robot remains a challenge, and, to date, researches hoping to succeed adopt three main approaches.

The first has proved its inefficiency at a general level, but should not be eliminated in some circumstances. We can call it, with Nelson (1991) "reconstructivism." It is based on the old idea of a universal robot and obeys the paradigm: using 2D images, let us reconstruct a 3D environment robot that will be able to understand thanks to Artificial Intelligence techniques. We know a lot of difficulties have never been overcome: environmental complexity, real-time, and mainly decisional processes that rule the whole system behaviour. Since a final and global solution is now abandoned, reconstructivism can be useful when drastic constraints in matter of environment and tasks are accepted.

The second approach is called "behaviourism" (and must not be confused with the well-known behaviourist philosophy). The idea is this: instead of directly creating an intelligent and autonomous robot, let us try to replicate the evolution of a baby as he progressively becomes aware of his body and learns to coordinate his motions with his sensory feelings. For that, a model of this evolution is necessary. As early as 1962, the famous psychologist Piaget (1962) proposed such a model based on the imitation phenomenon (of adults, then self imitation to improve coordination). No one knows if Piaget delivered a realistic model, but it is self-consistent and could be replicated with multisensory machines. Scientists hope, through this way of referring to the learning process, to bring a mechanical machine to true independence in its behaviour.

Finally, the third approach is called "reactivism" and was first proposed around 1985 by Brooks (Brooks and Mataric, 1993). The inspiration for reactivism draws its source from the observation of insects behaviour. One thing is certain with insects: their brains are so small and contain so few nervous cells that it is impossible for them to have decisional intelligence. However, they are able to intelligently transform their environment in order to survive and to transmit life to their descendants. Indeed, they act as automata, but every automation is guided by its interaction with the environment; that is, interaction occurs in a material animal-environment contact or in a remote contact with environmental attractors. So, reflex behaviour can drive them, individually or collectively, to perform a defined modification of the environment, that is to say, a task.

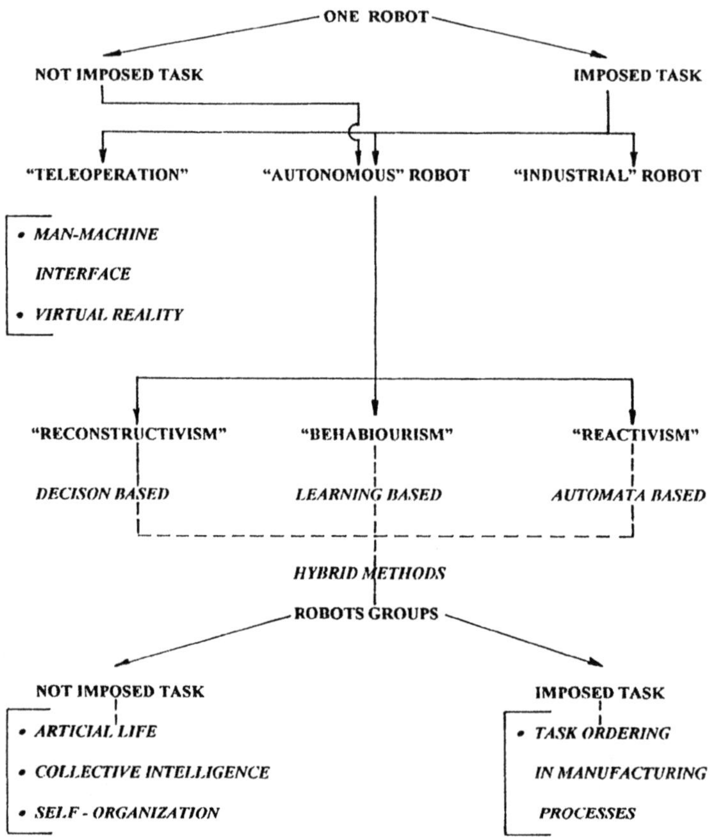

Figure 1.1. Scheme of the main up-to-date approaches to master robots.

Of these three approaches, two are recent and we cannot foresee whether they will be successful. We think, instead, that a hybrid using all the three approaches could be useful to master the autonomous robot problem.

Other researchers, instead of considering the "one robot or some robots" problem, work on behaviours of large groups of robots that are able to interact and cooperate. Again, we have to look at the case when the final task is prescribed (and the main application can be found in manufacturing problems dealing with tasks ordering) and to look at the case when tasks are not imposed. This is the favorite research of some computer scientists interested in artificial life, collective intelligence, or self-organization emergence. To date, few people pass from computer simulation to real robots.

The above overview (which I have summarized in the scheme) shows, in fact, two independent main pathways corresponding to whether man is or is not kept inside the control loops during the task performance.

This book is clearly located in the "Teleoperation and Man-Machine Interface" stream of Robotics.

As is well-known, teleoperation systems were first developed in the frame of nuclear activities to manipulate various dangerous items, including tools to machine these items, from which evolved the other generic name given to teleoperators - telemanipulators. Very early, people were aware of two main difficulties in using telemanipulators with efficiency: a lack of ergonomy in the master arm - slave arm (input device - executive organ) mechanical set; and, a weakness in the sensory feedbacks from the slave world towards the operator located in the master world.

This book addresses the first point, but also takes into account a part of the second point concerning feedback of forces that cannot be separated from the ergonomy problem. Indeed, the ergonomy problem is seen from an original point of view, which is that of quality analysis of the remote system.

Solutions to measure this quality and to improve it through specific control involving man-adaptive parameters summarizes the main topics. The proposals made by the author allow one to solve the everlasting problem of comparing various telemanipulators. They also establish a very good basis in the matter of control, allowing the addition of sophisticated systems that can be driven by powerful computer assistance while keeping the capacities of human dexterity.

The author is an "old timer" in the matter of telemanipulation, for he began to work on the subject in the former USSR as early as the beginning of the seventies. A bright and productive researcher, he met a difficult situation when he was obliged to leave his country as a consequence of the USSR history. I pay homage to his courage, and to his merit in succeeding to write this book after restarting his research from scratch in a new country in 1993. I hope, and am convinced, the favorable impact of this book on his scientific peers will suitable reward him.

References

Brooks, R. and Mataric, M. (1993) *Robot Learning*. Kluwer Academic Publishers, Boston.
France, A. (1927) *Les opinions de Jerome Coignard*. Gillimard, Paris.
Nelson, R. (1991) Vision as intelligent behaviour - An introduction to machine vision research at the University of Rochester, *Intern. J. of Computer Vision*, **Vol. 7 no. 1**, pp. 5-9.
Piaget, J. (1962) *Play, Dreams and Imitation in Childhood*. W. W. Norton, New York.

Philippe Coiffet
Director of Research at CNRS
1994 Award of Robotics Industries Association (USA)
Member of the Council for Application of the Academy of Sciences (France)

PREFACE

The telerobotics field is now highly developed and its literature is voluminous. While not ignoring this vast array of subjects, "Remote Manipulation Systems" focuses on problems of the remote control of manipulators in a free workspace. There is also considerable literature on this subject in a form of books, papers, and patents. This field deserves, in fact, not one, but a series of books on its different aspects, because the questions related to manual control of robots are very many and very complex. I have not set myself so comprehensive a target, nor do I have the intention of repeating material encompassed in well-known works.

Rather, I was concerned here with producing a book that would provide a survey of classic studies in telerobotics, and how their results relate to the principal subjects covered here. Corresponding papers and books are cited so the reader can obtain detailed background information from other sources, as well.

My main purpose in writing this book was to present relatively new data on remote control for telerobots, which is a subject that is poorly elucidated in literature accessible to a reader. Since the publication of the basic monographs of Vertut and Coiffet (1984, 1985) and Sheridan (1992), a number of new results have been obtained by both myself and by other researchers working, for instance, in Russia. This research and others will be illuminated in the book. The focus is concerned primarily with the influences of the basic properties of remote systems and how they affect their work quality. Remote control methods are presented in connection with two intimately associated problems of remote control quality: evaluation and remote control system design. The progression of this presentation includes:

- the evaluation of remote control quality;
- a proposal of means to improve the quality;
- development of corresponding schemes and devices in detail;
- developing methods of remote system analysis and design.

The work encompassed here has engaged no less than 25 years. It was begun in 1969, when I started my Doctorate in the Institute for the Study of Machines (Moscow, USSR Academy of Sciences). I continued in this field from graduation until 1991 at Kazakh State University (the former

USSR), where I worked in test stand implementation. I then began to focus on attempts to change the quality of remote systems by adjusting their parameters. A number of concrete systems were implemented and tested. Since 1993, I have been associated with Ben-Gurion University of the Negev, Israel, where I have participated in a human operator behavior study in remote control systems which is aimed at finding ways to predict the characteristics of these systems during the design process.

I hope that the data and ideas described in this book will be useful, not only in the remote control field, but also in the adjacent fields of robotics and automation. To that end, the book has been designed to be both informative in its field and as source material for researchers and engineers working in these fields and also graduate students with a corresponding specialization.

ACKNOWLEDGMENTS

I am very glad to have an opportunity to acknowledge some of the people who have supported not only my work on this book, but also my research over many years. First of all, I would like to remember the late Prof. A. Kobrinskii, whom I shall always consider my teacher. Special thanks are also due to Prof. P. Coiffet for supporting my different undertakings, including this book project. I greatly appreciate his suggestions for the manuscript, and for his generous offer to write the foreword. I wish to thank Prof. T. Sheridan for his recent, and very useful, collaboration with me, his review of the manuscript, and valuable comments.

Special acknowledgments go also to former colleagues from Kazakh State University, with whom many of the results discussed here were obtained. I would like to particularly distinguish Prof. U.A. Djoldasbekov for supporting our investigations and Dr. N.K. Djamalov for fruitful collaboration, beginning while he was a student, and continuing until he received his Doctorate. I am grateful also to my colleagues at Ben-Gurion University – especially, Profs. G. Ben-Dor and B-Z. Sandler, and Drs. Y. Edan and A. Melman – for support and encouragement during the period of writing this book. I especially thank Dr. Y. Edan for reading the book draft and her useful remarks. I would also like to thank Mr. M. Poizner, Dr. M. Dulger, and Mrs. O. Nirenberg, who helped prepare figures for the book, and especially C. Logan, for her skilled editing of the final text.

Finally, I am very grateful to my wife and children who almost never saw the husband and the father during the year while I prepared the book.

CHAPTER 1

INTRODUCTION

The present developmental level of modern techniques allows the automation of practically all technological operations, including those employing robots. A rather intelligent robot may now be created that can exclude the necessity for direct human operator (HO) control. On the other hand, there are always situations in practice that are not accounted for in robot software. Therefore, various remote control systems governed by a human have been explored. The HO has a number of shortcomings as a system control element. At the same time, he/she possesses unsurpassed abilities to recognize "situations" and to make rapid and usually adequate judgments to deal with them.

Moreover, even though we now have the technological know-how to solve any control problem by means of autonomous robotic systems, this does not mean that it is worth doing it. There is the factor of ***advisability***, which is frequently used as a principal argument when solving automatization problems of a certain technological process. Although this factor has not been very formalized yet, some of its more concrete components include:

1) Cost - the cost of a completely autonomous robotic system may considerably surpass that of corresponding remotely controlled systems.

2) Labour consumption of design and manufacturing - it is clear that autonomous systems are more complicated to design and make, which means the process can be longer and can delay its practical implementation.

3) Exploitation complexity - this index may also be presented in a money form, because an autonomous system requires great expense to maintain its equipment.

4) Reliability - a non-zero probability of failures in an autonomous system often forces its backup to perform with the participation of the HO.

This fourth point suggests that, if it is already necessary to frequently employ the backup of the automatic system with the help of manual control, it is worth designing a system that includes this option at a high performance level in order to increase the overall performance of robotic systems. Thus, telerobotic systems have been developed. Current requirements of different kinds of industries have very much stimulated development in this direction because sometimes there is simply no alternative to the telecontrol. Because of this, telerobotics is a field where new and very interesting results constantly appear. New problems in the practical application of robots are constantly arising, and many of these problems should be solvable by telerobotics. For this reason, the field of telerobotics is developing much more rapidly than industrial robotics, whose circle of problems is more stable. O'Hara and Olsen (1988) stated that a space, for instance, offers a challenge to robotics technology because few operations in space are routine. The important conclusion is that real operation technological requirements are somewhat beyond the capability of fully autonomous systems. This is, in fact, the whys and wherefores of remote handling technology application.

Additional advantages of this approach is that these systems may not only take the place of humans when it is necessary to work in hazardous environments, but they also ensure fulfillment of such operations that cannot be performed by a human without the use of auxiliary techniques, such as those needed to exceed human power, accuracy, arm size, etc.

HO participation in telerobot control may take place at different hierarchical levels (Fig. 1.1), such as an actuator level, an end effector level, and object and objective levels (Vertut and Coiffet, 1985). These are differences in the levels of interaction between the HO and the robot, which usually depends on the degree of robot development. The interaction intensity between the HO and a robot also varies according to robot types.

Remote manipulation systems are, in fact, some part of telerobotic systems. We consider here the remote manipulation systems as a mean to perform control operations in telerobotic systems, usually on their tactical level. These systems also include a number of other subsystems; for instance:
- the means of an executing level to control a telerobot mechanical arm (Medvedev *et al.*, 1978);
- the means to obtain and present to the HO different kinds of information (primarily, visual information, but also force information, etc.);
- the means to obtain this kind of information to assist computerized decision-making (Vertut and Coiffet, 1985);
- the means to effect the HO interaction with a computer (including high-level languages) (Sheridan, 1992);
- a prediction subsystem (Hirzinger, 1993).

The range of problems with which telerobotics deal also includes the ergonomic design of human control and information means. All these subjects are presented in this book briefly and only in the context of our principal subject.

The multitude of telerobotics areas require different approaches. For instance, one approach demonstrated in Sheridan's book (1992) was based on human engineering and ergonomics. An approach closer to the author's

HIERARCHICAL LEVEL			
I	**II**	**III**	**IV**
Executing	**Tactical**	**Strategic**	**Highest (intellectual)**
Motion control in each degree of freedom	Distribution of control signals at degrees of freedom	Operation separation for elementary constituents	Situation recognizing and decision making about operation

Figure 1.1. Different hierarchical levels of HO interactions with telerobots.

was that of Vertut and Coiffet (1985), which was directed toward describing technical problems of telecontrol. However, these works and other published papers have not identified or discussed a number of technical questions of remote control.

An important problem is that of optimal coordination of the HO with a robot or a teleoperated system under manual control. Such integration of robots with humans has been a very thorny problem in telerobotics, and its great complexity still defies definitive solutions.

To get an idea of this complexity we can look at Figure 1.2, which summarizes the general problem by showing the ways of transmission for "actions" and "information" (Slutski and Coiffet, 1996). In most cases, action is supported via forces and torques (generating reactions). Information feedback can be visual or haptic. The master-robot link is seen here in terms of control assuming all other ways as perfect, with the aim of improving the quality of the entire system.

In this work we assume all forward and backward edges of the scheme as corresponding to a heuristically correct behavior of the system, with

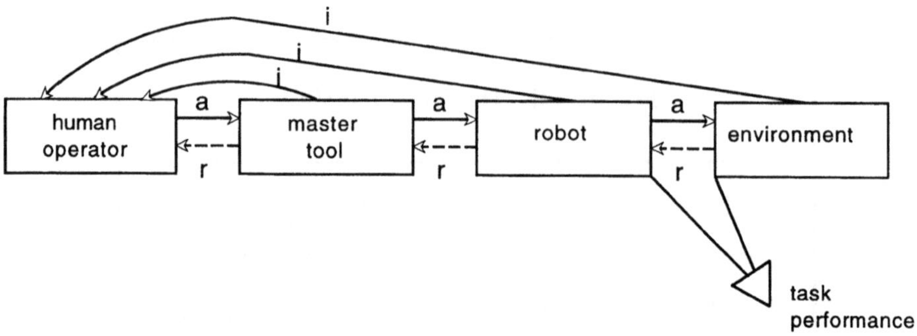

Figure 1.2. General scheme of manual control of robot: *a* - action and reaction ways, *i* - information feedback ways. From (Slutski and Coiffet, 1996). ©Cambridge University Press.

the exception of the "master-robot" relations, for which we propose more attentive study. Also included is a discussion of improvement of this link using new control techniques in order to facilitate the user friendly behavior of the system while being piloted by the HO.

The present-day control theory is one of the most powerful tools for the improvement of robot performances. When applied to the field of remote control, it allows one to solve most of the practical problems in robotics, but not always by using the most effective method. One of the cardinal problems of telerobotics is that, according to results of various studies (e.g., (Kobrinskii *et al.*, 1971)), the two principal indices of these systems - accuracy (error) and task fulfillment time - are connected by inverse proportion (Fig. 1.3) for a coordinate position system; that is, the criteria are contradictory. Thus, by improving one of these criteria, the other is decreased. In Figure 1.3, the dependence is the Pareto set, which is utilized to choose some optimal solution. As usual in this situation, an additional criterion has to be introduced to determine the solution. By solving a design problem of the system in this way, however, we cannot obtain a flexible solution that ensures both criteria.

Non-standard approaches are thus required to solve non-standard problems. One such approach is the adaptive coordinate-parameter control, which permits one to solve complicated, multi-purpose problems when the main control problem combines with additional control conditions. In particular, certain optimality requirements may be present as in such conditions. The parameter control channel allows one to realize these requirements at the same time as the main purpose of the control process is achieved. Frequently, this solution may be obtained by a simpler approach and with the help of simpler technical means.

"Remote Manipulation Systems" is based on a kinematic approach to re-

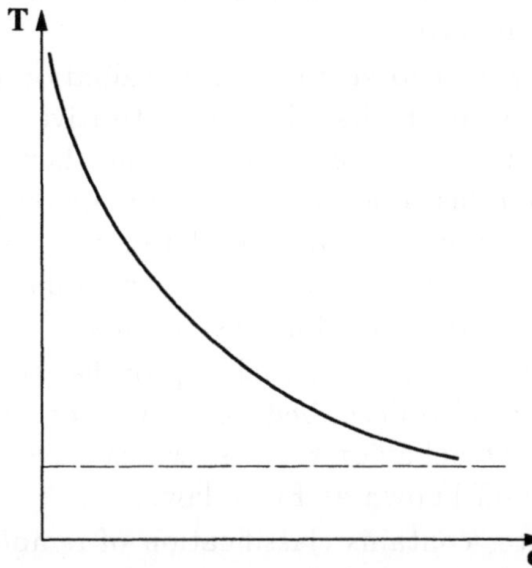

Figure 1.3. Trade-off between telerobot positioning error Φ and task fulfillment time T. From (Slutski, 1997). ©1997 by the American Nuclear Society, La Grande Park, Illinois.

mote manipulator control. Therefore, the consideration of the force control that contributes to control quality increase is omitted, but not forgotten. The author believes that this question goes far beyond the scope of the current subject, which is to deal with an important type of remote manipulators that work with unconstrained objects, which is, in fact, the basis for any remote system. If this system can maintain a high level of indices and allows effective solutions to basic manipulation problems, it may be added to further by other devices (for instance, force reflection, prediction system) to increase its work quality. Thus, such systems are a major factor in influencing remote systems quality.

Concerning terminology, types of control systems have various names, and some of these will be used in the book. In addition to the primary classification ("remote manipulation systems"), we will use also the terms "telecontrol systems," and "manual control systems," as well as "man-machine systems" (relating only to their control processes). We also use the term "man-machine systems" because of habit over many years, even though the term "human-machine systems" is more general.

A short survey of the book content is as follows:

In the **second** chapter, the data on purposes, structures, and kinds of remote manipulators are presented. This short chapter encompasses materials that are known to specialists in the field. The most important point relates to control methods for present-day telerobots. It is pointed out that such control is always accomplished with the help of the well-known coordinate approach; e.g., on the tactical level, the control of the generalized coordinates of a manipulator.

The **third** chapter is devoted to quality evaluation of remote manipulation systems. An approach, based on quantitative resultant criteria, is delineated. It is based on the fact that, all other factors being equal, the criteria of accuracy and fast-operation are the main output indices of these systems. Applications under conditions of the remote control simulation and actual manipulator checking are shown (manipulator checking is implemented by means of corresponding test stands).

Analytical-calculate computer-based approaches to analyze and design telerobotic systems are also described. The more promising approaches of these are based on HO behavior studies, which have their origins in the speed-accuracy trade-off known as Fitts' law.

The **fourth** chapter contains classification of remote manipulator control methods and their detailed description from the viewpoint of their ability to meet the quality requirements of a manipulation system. This chapter also offers a detailed discussion of universal input device for the master-slave manipulator based on a flexible working element. It has been shown that high-quality remote control systems of the tactical level are now too complicated and always entail high cost and low reliability, which means that further improvement of existing remote control methods is required.

The **fifth** chapter sets out an approach to the development of a new class of telerobots with variable parameters. An experimental investigation, performed with the help of semi-natural simulation, served as the basis of this approach. Its results showed that the most influential parameter for all system criteria is the manipulator gain; therefore, if there is a need to accomplish telerobotic work in an optimal regime, the gain is the first parameter that should be adjusted. A solution is proposed and described where the operator himself adjusts the manipulator parameters during the control process according to requirements of a specific situation. An additional channel of parameter control that acts in conjunction with the main coordinate channel is then introduced into the system. This approach results in the organization of semiautomatic adaptive control systems.

The basis of the remote manipulator adaptive control is laid down in the **sixth** chapter. Classification of such systems is presented, where all the main kinds of systems are described. The grounds for determining man-adaptive algorithms are stated and integral-type algorithms are synthesized. Other versions of the control systems are also considered.

Results of the simulation of this type of system under semi-natural experimental conditions are described. A significant increase in telecontrol quality under adaptive method usage is noted. The varieties of real industrial robot remote control systems, based on the adaptive method, are also described. Their test results show that control time decreases by several magnitudes for a given accuracy level, as compared to ordinary position control.

Another approach to adaptive control utilization in manipulation systems is described in chapter **seven**. This uses mechanical means for the system gain control, and may be used if, as stated in the previous chapter, it is not possible to vary the gain in the manipulator control system. Thus, there is still a possibility of changing the manipulator's mechanical scale coefficient. This coefficient, in its turn, may be controlled by one of two methods. The first solution is associated with scale alteration of the manipulator slave arm. The manipulator construction is then somewhat complicated, but the achievable effect is not limited by the advantages of the adaptive control. As this take place, manipulator accessible region volume, maneuverability, and other characteristics are increased, while the non-utilizable area volume is decreased. The problem in hand can also be solved by changing the scale of the master arm, an approach which is simpler and cheaper.

The **eighth** chapter proposes methodologies to evaluate and predict telerobot performance. The first of these methodologies is based on an adaptive model of HO behavior. This approach allows us to evaluate quality characteristics of remote systems with different kinds of machines. Its feature is that although it performs a rather approximate, preliminary estimate of these characteristics, it enables us to achieve principal analysis and synthesis of these systems, corresponding with remote system requirements of accuracy and rapid-operation.

Another methodology is based on the study of HO motor behavior in making discrete movements. The analysis of the experimentally obtained operator motions revealed a control process structure as a consequence of individual, partially superimposed submovements. The established motion structure enabled an estimation of the remote control process duration. A specific performance evaluation was applied to estimate the characteristics of different remote control methods, including a developed adaptive algorithm. The obtained results were used to predict duration characteristics of new remote adaptive control algorithms. This approach can be used for the preliminary design of remote control systems in order to limit experimental investigations.

References

Hirzinger, G. (1993) Multisensory shared autonomy and telesensor programming - Key issues in space robotics, *Robotics and Autonomous Systems*, **11**, pp. 141-162.

Kobrinskii, A.E., Sergeev, V.I., Slutski, L.I., and Tyves, L.I. (1971) On the quality evaluation of manual control systems, *Mashinovedenie*, **no. 5**, pp. 3-10 (in Russian).

Medvedev, V.S., Leskov, A.G., and Yushchenko, A.S. (1978) *Control Systems of Manipulation Robots.* Nauka, Moscow (in Russian).

O'Hara, J.M. and Olsen, R.E. (1988) Control device effects on telerobotic manipulator operations, *Robotics and Autonomous Systems*, **Vol. 4 no. 1**, pp. 5-18.

Sheridan, T.B. (1992) *Telerobotics, Automation, and Human Supervisory Control.* The MIT Press, Cambridge, MA.

Slutski, L. (1997) Adaptive control techniques for teleoperated manipulators, *Proc. ANS 7th Topical Meeting on Robotics and Remote Systems*, Augusta, Georgia, pp. 661-667.

Slutski, L. and Coiffet, P. (1996) Man-adaptive control of teleoperators, *Robotica*, **Vol. 14**, pp. 111-117.

Vertut, J. and Coiffet, P. (1985) *Teleoperation and Robotics: Applications and Technology*, Vol. 3B. Prentice-Hall, Englewood Cliffs, NJ.

CHAPTER 2

REMOTELY CONTROLLED MANIPULATORS: USE AND STRUCTURE

2.1. Purposes and Applications of Remote Manipulators

Among all robotic devices, manually controlled manipulators have a special place. Remote manipulators were the first variety of manipulators to appear in the 1950s. The need to develop remote controls happened when radioactive materials had to be handled in industrial situations. Remote manipulators have also been found to be useful in many highly complicated and responsible tasks that must be performed in "extreme environments." These tasks have become some of the most important with which modern robotics deal with. They are performed in space, undersea, in the context of nuclear power engineering, and in military applications, where it is difficult (or impossible) to ensure a satisfactory technical result without HO availability in control circuits. In presenting these applications, we do not compete with Vertut and Coiffet (1985), whose book is highly recommended for its great wealth of details. We will limit ourselves only to a brief enumeration of the uses of telerobots and move on to the presentation of relatively new data.

It is possible to divide all known robotic applications of these systems into two quite different groups: non-industrial and industrial applications.

Concerning to the first of these groups, note that this includes intensive research in space, underwater (Anderson, 1989; O'Hara and Olsen, 1988; Ollend, 1989), robotics for nuclear industry (Andreenko *et al.*, 1975; Herndon and Hamel, 1987), etc. These robots are sometimes called "service" robots.

Space telerobots allow astronauts to work in a comfortable environment, while acquiring and manipulating payloads, such as satellites, space station modules, and orbital replacement units. These tasks are performed by means of telerobots like that shown in Figure 2.1, whose manipulator is, in fact, some combination of a large arm with a great workspace and an attached, relatively small, dexterous arm for fine operations.

Underwater telerobots may be used to perform a number of different functions. Vertut and Coiffet (1985) cite the most important as:

- sinking and embedding cables and pipelines; and
- undersea mining.

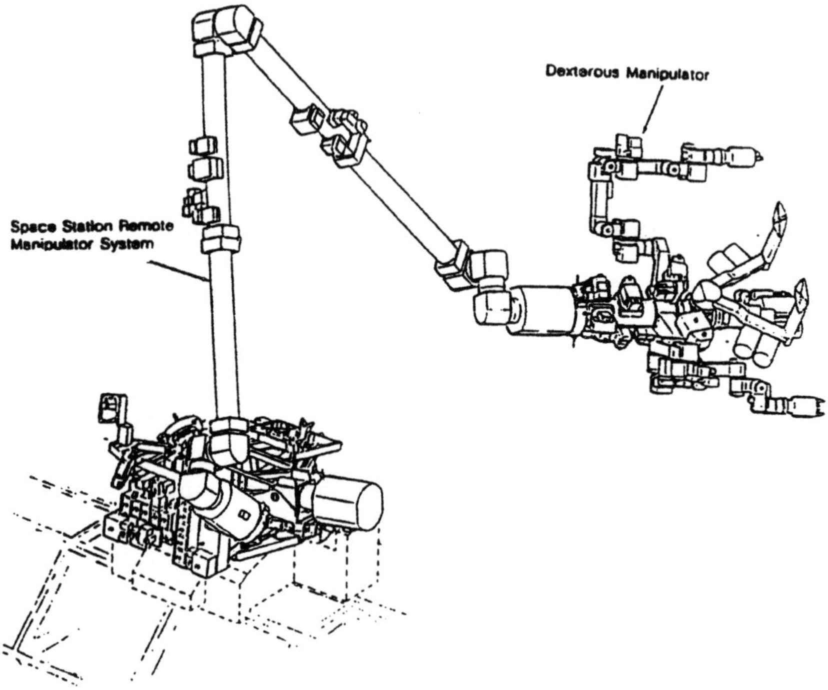

Figure 2.1. Space shuttle remote manipulator system. From (Brimley *et al.*, 1994); ©American Institute of Aeronautics and Astronautics.

The latter task is closely related to offshore oil exploration, where remotely operated vehicles with manipulators are able to build and inspect undersea platforms.

The space and underwater robotics have a number of common problems because of the similar kinds of constraints they place on HOs. There are two principal approaches to organizing these systems. The first is based on the use of some kind of ground-control, and the second is to control the systems within direct proximity through space or underwater apparatus. For the space applications, the first approach is potentially much more effective (Ruoff, 1994) for a number of reasons:
- the total hourly cost of a ground-based operator is orders of magnitude less than that of an operator in a space;
- the HO works in normal environments (for instance, without any geometrical restrictions), which increases work quality;
- a ground control station has greater computing resources, better possibilities to maintain itself, and may be configured to support many operators;
- risk to HOs is decreased and even suppressed.

Figure 2.2. **The mobile telerobot TSR-700 for complex, high-risk military and civilian tasks. Courtesy of TAAS-Israel Industries Ltd., Israel.**

On the other hand, on-board control does not have the most important disadvantage of ground control: communications time delay, which is the reason for many problems encountered in space operations. Control system instability may, for instance, take place when a telerobot becomes inefficient. We do not deal here with this important part of modern telerobotics, and instead refer the reader to Sheridan's book (1992), which presents this problem very fully.

Concerning on-board control, we note that it has its own disadvantages:
- a restricted working area, which prevents HO motions and decreases control possibilities;
- limitations on control and mapping device numbers and weight, which leads to the possible reduction of control quality, etc.

In practice, a designer has to choose some "golden mean" between these two solutions by combining them to build an effective remote control system. Present-day space telerobotics has achieved very great progress, but many problems have not yet been solved. According to (Ruoff, 1994), there is sometimes far less astronaut time available than is needed to perform necessary inspections and maintenance operations.

Further, it is necessary to note military applications of telerobots. Bomb disarming and ammunition disposal problems, for instance, are related to this class of teleoperations. An example of these telerobots is presented in Figure 2.2. This mobile robot has independent, four-wheel drives, which allows it to moving the most difficult wet or dry terrain, and even to climb stairs. Its hydraulic arm has seven DOFs, with an arm length of 2.85 m and a lifting capability to 140 kg. In addition, it may be modified to perform nuclear and chemical material detection, decontamination, fire fighting, take part in anti-terrorist activity, etc. Thus, it can be seen that special robot development for the hazardous nuclear, chemical, and biological environments of the modern battlefield will serve to advance present-day telerobotics. We can hope, however, that these robots will not ever have to work in such conditions.

A highly developed application of telerobotics is in the medical use of robots; e.g., for surgery and rehabilitation. Note that the trend of development in robotic surgery mostly involves methods of remote control. One of the purposes of telesurgery is to make it possible for expert surgeons at regional medical centers to actually participate in surgeries at local clinics; for instance, (Green *et al.*, 1995) describes an approach to provide surgical services in combat zones.

In the context of nuclear technology, remote manipulators also perform different kinds of very responsible operations. This field is, in fact, on the frontier between industrial and non-industrial applications.

A radiation hazard was the first factor that promoted the appearance, not only of telerobotics, but also of general robotics. There are many possible robot applications in nuclear power plants. In addition to the traditional field of radioactive maintenance, mentioned by Vertut and Coiffet (1985), Roman (1989) lists the following kinds of tasks:

- radiation monitoring/smear tests;
- sampling radioactive materials;
- small spill cleanup;
- changing filters;
- steam generator and condenser tube sheet repair.

Other literature sources point out additional hazardous jobs:

- removing radioactive waste stored in underground tanks;
- work in nuclear reprocessing plants and hot cells;
- cleanup of chemically and radiologically hazardous waste;
- decommissioning (or demantling) of nuclear plants.

Among the qualities demanded of these manipulators, ease of control is prominent (Andreenko *et al.*, 1975). This property is important because a limited field of vision and HO restricted motions cause rapidly-building fatigue and consequent errors. To reduce these problems, a design of telerobot input devices (IDs) with minimal weight has been proposed. Other goals of design should be accuracy and sensitivity.

Telerobots are also used for tasks involving chemical (toxic), explosive, and unexploded materials, which are similar to telerobot use in the nuclear

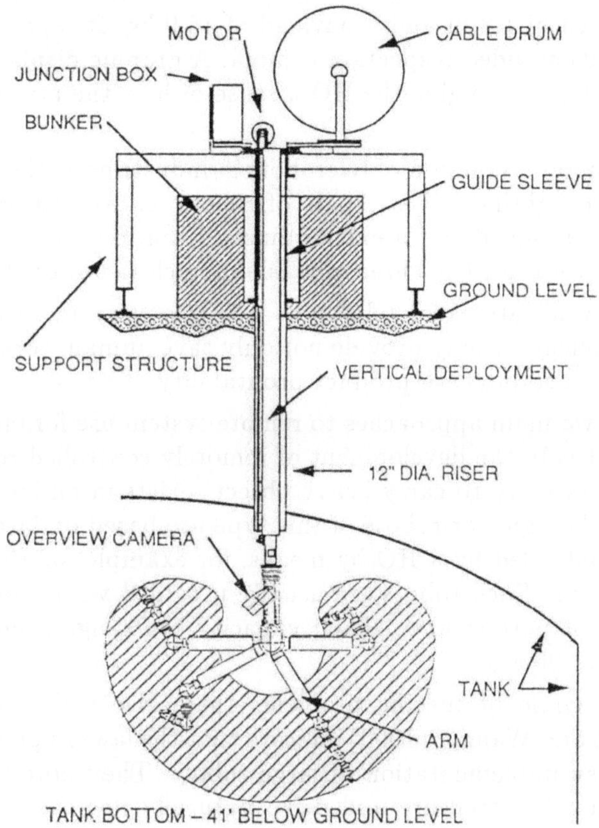

Figure 2.3. Manipulator for remote tank inspection. From (Knape and Berger, 1991).

industry.

In these various fields of telerobot application, it is possible to identify common problems to be solved. First of all, inspection problems arise in all fields, without exception. In space robotics, it is necessary to check, e.g., the state of the surfaces of satellites and space platforms, the booster rockets of a shuttle payload cradle, and truss struts for micrometeoroid damage (information of Jet Propulsion Laboratory). These inspections are performed by a robotic arm, which incorporates a television camera.

In a nuclear power station, inspections of the steam generator, reactor containment, internal/external piping, storage tanks, etc are required. An example of a remote tank inspection robotic system is presented in Figure 2.3 (Knape and Berger, 1991), which performs remote visual inspection of corrosion inside high level liquid waste storage tanks. The system manipulator provides 5.8 m of linear extension inside the tank to position a 5-DOF arm with a reach of 1.8 m and a payload of 15.9 kg. Its end effector (EE) is a high resolution video inspection camera. A graphic display is provided at the control console to give the HO a sense of how the arm is positioned inside the tank.

Another common function of telerobots in industries as these mentioned above, is in fire fighting. These kinds of robots, as well as other types of robots that are designed act in extraordinary situations, were development to high levels after the 1979 Three Mile Island nuclear power plant accident.

Industrial applications of telerobots are also very important. In this field, robot application may provide not only save human workers time and effort, but may also increase product profitability.

There are two main approaches to remote system use for industrial purposes. The first is in the development of remotely controlled robots of high payload. The necessity to carry heavy objects exists in all industries without exception. Designs for robots of this type are based on large hydraulic-driven arms controlled by a HO by means, for example, of a small copy of the executing arm. Such robots are usually provided with a force feedback to improve robot performance. Their capacity can range from around 250 kg to more than 2000 kg.

There is a circle of specific problems that arise while creating large robots. Schraft and Wanner (1993) indicate that the laws of physics mitigate against the easy implementation of large robots. They note that as robot becomes larger, the structure and drive naturally decreases, limiting the system gain and rapidly degrading the accuracy and speed of operation to an impractical level. Thus, the large robot design is a difficult problem, although it has been successfully solved by several research teams.

A typical example was the manipulator "Andromat," which was produced by the CSEE Company. We present here another example of such robots, which was primarily intended for tasks in foundries (Fig. 2.4).

Other uses of power telerobots are related to the performance of some technological operations. One of these, an application for performing remote mining processes, was described by Thring (1983).

The electrical industry is also potential area for robot use. Roman (1989) identifies two possible domains for this: fossil power plants, and sub-station/switching stations. In the latter case, examples of teleoperated problems might include:
- operating energized equipment;
- cleaning transformer coolers;
- opening energized disconnect switches;
- operating test equipment.

Figure 2.4. Hydraulic manipulator "Scobotman" for heavy-duty operation. Machine capacity: 250 kg; Components being handled: Cast iron brake drums and bearing caps; Weights of components: Up to 35 kg; Production rate: Up to 1440 castings/hour. Courtesy of James Howden and Co.

One of these tasks is the maintainence of electric power distribution aerial lines (Goldenberg *et al.*, 1995); robots carry out such operations as exchanging porcelain insulators, replacing conductor aerial supports, and cutting tree branches that could damage power lines during storms (Fig. 2.5).

Another example of telerobot use in this area is their application in forestry operations. Here there is a major and important operation where

Figure 2.5. Remote manipulator for forestry operations which consists of a robot arm (treejib) to be mounted at the tip of the aerial device. Courtesy of A. Goldenberg, University of Toronto, Canada.

telerobots may be effectively used: felling trees. In this case, a power robot clutches a tree near its base and saws through it with the help of an EE in the saw. It may also perform delimbing, and afterward, lay trees or logs on a special vehicle. These robots are an effective mean for removing humans from work that is too hard and dangerous.

A characteristic example of large service robots is the aircraft cleaning robot "Skywash" (Schraft and Wanner, 1993). The creators of this robot had to consider the following:
- the outward appearance of a plane is related by the passenger to the standard of the airline;
- corrosion of the plane can be retarded by regular cleaning.

For these purposes, a robot was designed which consists of a large,

Figure 2.6. A general view of the aircraft cleaning robot Skywash. Courtesy of R.D. Schraft, Fraunhofer Institute for Production and Automation, Stuttgart, Germany.

multisection boom with six main axes, three wrist axes, and two adaption axes (altogether 11 programmable axes). A brush is located on the robot end as its EE. Hydraulic drive systems were used for all these axes. Two adaption axes are an original part of the arm structure, and they ensure fine control based on the measured pressure of the brush applied on the plane surface.

A general view of a developed robot is shown in Figure 2.6, while Figure 2.7 shows the robot as it washes an aircraft surface.

All movements of the manipulator are generated off-line. However, a very important calibration procedure (which allows the separation of errors of the position measuring system and other kinds of errors), as well as some other typical control operations, are carried out by the HO from a remote control panel.

In 1993, the Skywash robot was tested during a six-month period at Frankfurt Airport on several B747 and B737 aircrafts. Among the benefits obtained when using the robot were:

- reduction in manpower per washing event by a factor of eight for wide-bodied planes and by a factor of four for narrow-bodied planes (e.g., B747 - 95 man-hours manually against 12 man-hours with the Skywash robot);
- reduction in ground time per washing event by a factor of three for wide-bodied planes and a factor of two for narrow-bodied planes (e.g., B747 - more that nine hours manually against three hours with two

Figure 2.7. The Skywash robot during a washing process. Courtesy of R.D. Schraft, Fraunhofer Institute for Production and Automation, Stuttgart, Germany.

Skywash robots).

The second important application of such systems is their use as a teaching method for industrial robot (IR) programming. The IR "Kuka" is shown in Figure 2.8 as an example of a highly advanced machine of this type that uses the mode of remote teaching. This mode implementation will be described below (Chapter 4).

There are two method versions in general use. For the first version, the HO moves a robot through a necessary array of configurations by using a special handle placed near the robot EE. This method, called "lead-through teaching," will not be discussed here for two reasons. First, this method is difficult for the HO to do, it does not give a good accuracy, and it requires a special techniques for its implementation. Second, this approach is not actually a remote control method because the HO catches, literally, a robot hand while he/she controls it.

Another approach to teach robots is based on the use of a special device called a teach box or teach pendant. In this case, a real remote control process has place. Nearly 90% of IRs have been programmed with the help of this method, according to Schreiber (1984). In recent years, this figure has slightly decreased (70% in 1993 (Hirchfeld *et al.*, 1993) and 63% in 1995 (Morley and Syan, 1995)). In spite of this, Morley and Syan believe that teach pendants are still one of the principal means of programming a robot

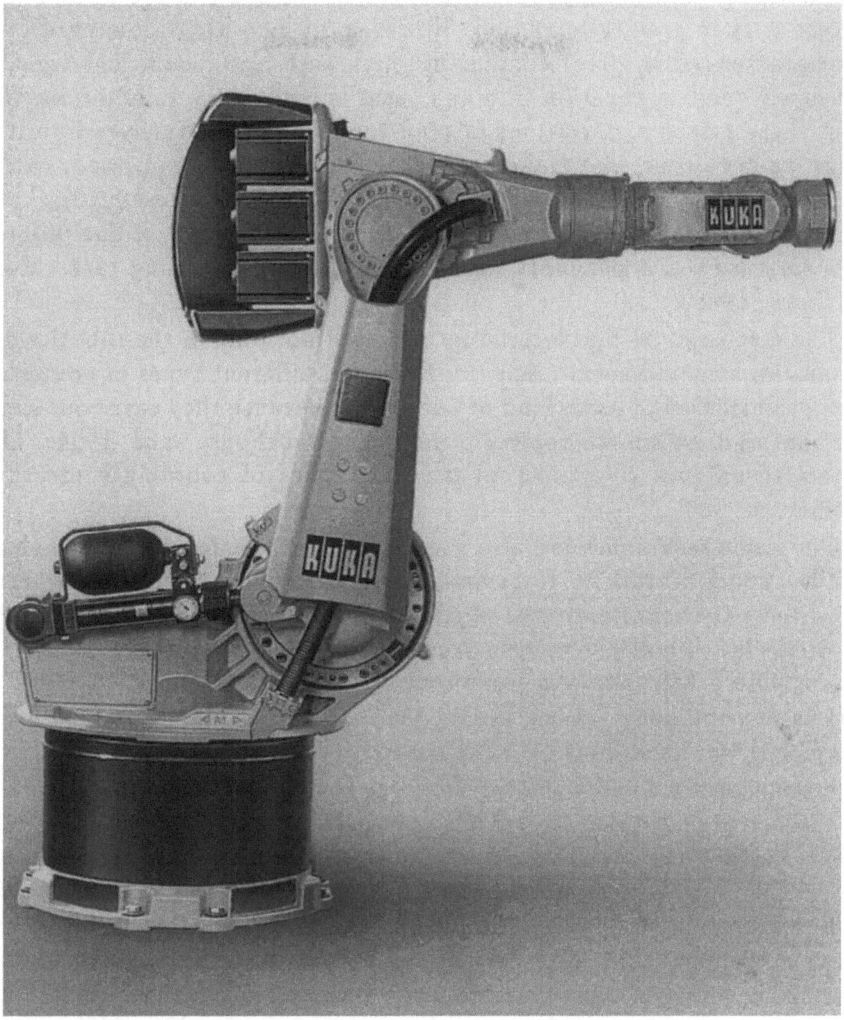

Figure 2.8. Industrial robot "KUKA," model KR 150/1. Payload: 150 kg; Number of axes: 6; Repeatability: ±0.2 mm; Maximum reach: 2410 mm. Courtesy of KUKA Roboter GmbH.

and that their use is likely to continue for many years.

Because of all its disadvantages, however, (see Chapter 3), this kind of programming in an on-line regime causes a bottleneck in robotized manufacturing systems. Overall IR utilization is reduced, since the IR is removed from the production process for lengthy periods (Langner, 1986; Schwartz, 1985). The drawbacks of teaching IRs by on-line programming serve the rationale to create a methodology for IR autonomous or off-line programming (Cardaun, 1990). The effective use of IR in the technological process is then greatly increased. However, off-line programming systems are characterized by great complexity, high cost, and insufficient versatility; hence their application in many cases is ineffective. In addition, when using these systems, deviations of true values of IR parameters from their design values occur, and therefore additional, rather complicated, calibration systems for IR (Judd and Knasinski, 1987; Ziegert and Datseris, 1989) are necessary. However, it is important to note that many off-line programmers also use teach pendants to complete the programming task (Morley and Syan, 1995).

The discussed on-line techniques are used not only in the robotics field, but also in some adjacent fields; for example, different types of cranes may also be considered as some kind of telerobots, because they carry out similar functions and use similar control techniques (Backhouse *et al.*, 1994). These control techniques is applied to manual modes of coordinate measuring machines as well.

The same techniques are also used to control the docking of the shuttle with the space station (O'Hara and Olsen, 1988). It is the station operator who directs the remote process of six-DOF targeting an EE alignment with a grapple lug. Finally, remotely controlled movable vehicles also use such means. These vehicles - e.g., different rovers - are intended to move and work in environments on the Earth, Moon, or other planets, where human presence is too hazardous or, simply, impossible.

It is necessary to note that present-day remote control systems work not only in a real environment, but also in a virtual one (Burdea and Coiffet, 1994). As they are special tools designed to allow input to the computer, they ensure in this case human-computer interaction (Chen *et al.*, 1995).

In the Virtual Reality environment, an important application of these means is, in particular, as a part of the hardware to improve the operator interface in modern CAD systems. There, these serve to perform the most elementary operations of, for example, pointing and tracking while creating and modifying a component model (Springer and Gadh, 1996). The use of advanced remote control techniques in this case allows one to intensify the design process and, finally, to increase the efficiency of the CAD systems.

2.2. Varieties of Remote Manipulators

Manipulation robots that include in their structure a remote control system may be divided into several groups.

1. Proper remote manipulators

The robots related to this group were already mentioned in terms of robots that are designed to work in extreme environments and those that have high load-capacity manipulators. In this case, the HO constantly interacts with a telerobot by controlling it.

2. Industrial robots

IRs are among the most important applications of remote control systems. Remote control systems are used to program IRs by teaching them; that is, the HO teaches a robot, and then the robot performs the received program automatically.

3. Supervisor control systems

These systems have maximum use of the possibilities of both automatic and remote controls of manipulators (Sheridan, 1992). Briefly, such systems work in two main regimes:

I - An automatic control regime. In this case, the system carries out a task in an automatic, autonomous regime based on robot hardware and software;

II - A remote control regime. This regime is used in cases when non-stationary environments make it impossible for a robot to work under an automatic regime. The reason for this is that a robot's software and hardware may not be adequate to deal with new and unexpected work conditions.

At the latter case, a control is switched to a remote mode either automatically or by the HO him/herself.

It is obvious that such non-stationarity is the most usual for non-industrial applications of robots. These primarily relate to robot work in unstructured conditions (space, undersea, weapons ...), where non-stationarities are an integral part of a robot's environments.

While working with this system, the HO can perform control on two different levels: tactical and strategic. On the tactical level, he/she sets remote control signals directly from a hand controller (HC). On the strategic level, HO activity consists of forming commands that enter a high-level computer-supervisor.

By working with such a system, the HO performs four principal functions:

- planning (this function of strategic levels is finished by producing commands for the tactical control level);

- teaching; that is, introducing commands to the computer-supervisor which are obtained by planning;

- observation; that is, obtaining information about telerobot actions in order to produce subsequent control plans;

- manual (remote) control regime; that is, direct robot control with the help of HC.

Another variety of a modern robotic remote control - shared control - is close to the supervisor control. In this case, however, the HO produces

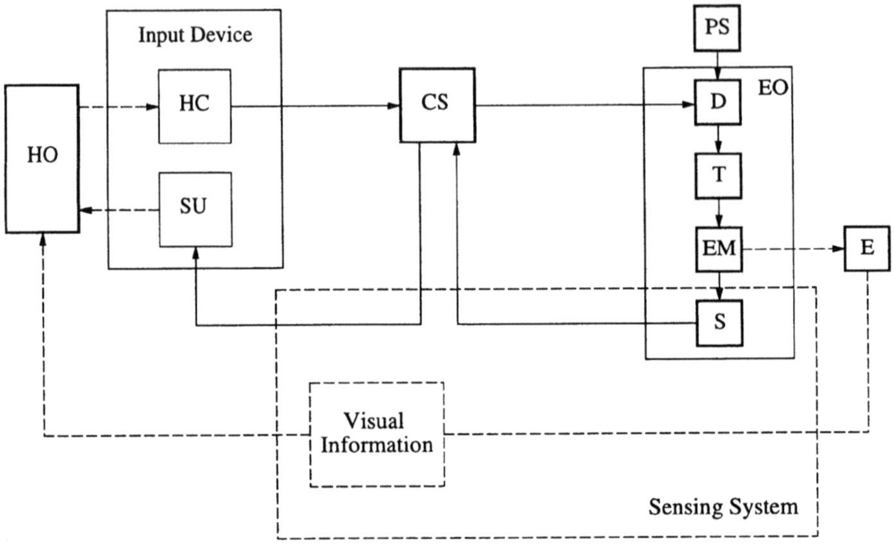

Figure 2.9. Block diagram of a remotely controlled manipulator. Designations: HC is a hand controller, CS is a control system, PS is a power source, EO is an executing organ, D is drives, T is motion transmissions, EM is an executing mechanism, and S is a sensor system.

control signals for some telerobot DOFs at the same time as those for other DOFs are produced by a computer. Thus, remote control action and automatic control are realized, while using the shared control, not in turn, as it was for supervisor systems, but in parallel, for different groups of robot drives.

So, it may be seen that almost all types of modern robots incorporate means of remote control.

2.3. Structures of Remote Manipulators

A block diagram of a remote control system is presented in Figure 2.9. Executing organ is usually a manipulator. The executing mechanism interacts with environment, and corresponding information enters the HO through a sensing system. The sensing system generally includes visualization means (for example, technical vision system), as well as other kinds of information channels (e.g., acoustic, haptic).

The CS (Fig. 2.9) consists, in its turn, of a number of functional blocks, and is usually implemented as a computing device with corresponding software. During kinematic control with force reflection, for instance, at least three blocks may be contained in its structure:

1) a sub-block of the actual remote control (SBC) that implements control algorithms acceptable for the type of control method being used;

2) sub-blocks (SBT) which carry out kinematic transformations of tele-robot frames. In general - for example, for position control - three kinds of these transformations may be performed:

- transformation of a vector $\vec{\phi}(t)$ of ID joint coordinates into a vector $\vec{r}(t)$ of components of Cartesian coordinates (a direct ID model);

- transformation of a vector $\vec{R}(t)$ of manipulator gripper Cartesian coordinates into EO joint coordinates $\vec{\Theta}(t)$ (an inverse robot model);

- transformation of an actual vector $\vec{\Theta}^*(t)$ of manipulator joint coordinates into a vector $\vec{R}^*(t)$ of its gripper Cartesian coordinates (a direct robot model).

Functional blocks realizing these functions may be labeled as SBT1, SBT2 and SBT3, respectively.

3) Finally, there is a sub-block SBS in the CS structure that transforms force sensing signals entering from the robot EO. After processing, signals $\vec{p}(t)$ from its outputs enter inputs of a sensing unit (SU) of ID (Fig. 2.9).

Based on introduced designations, a general block diagram of the CS for position control with force reflection is presented in Figure 2.10. Such systems are also called bilateral systems, because they feed back to the HO signals about robot interactions with its environment. These may also be unilateral if control signals go only to one direction from the HO to EO; that is, if force feedback is not in use.

In the block diagram shown in Figure 2.10, note that transformations of sub-blocks SBT1 and SBT3 present solutions of forward kinematics, while sub-block SBT2 solves inverse kinematics. The solution quality essentially influence all robotic systems quality (not only manually controlled, but also automated systems); therefore, many studies are devoted to this problem (e.g., (Asada and Slotine, 1986; Craig, 1986; Fu *et al.*, 1987)).

While analyzing the block CS structure, note that algorithms of sub-block SBC have great importance for system performances. The analysis and development of these algorithms are therefore one of the main purposes of our work.

We would like to note that, on the tactical level of modern manipulation robots, control is always accomplished with the help of a coordinate approach; that is, to say, the control of generalized coordinates of a manipulator mechanical arm (Whitney, 1972).

2.4. Conclusion

In this chapter, a brief survey of robot applications, varieties and structures is gone. It shows that on-line remote control methods occupy an important

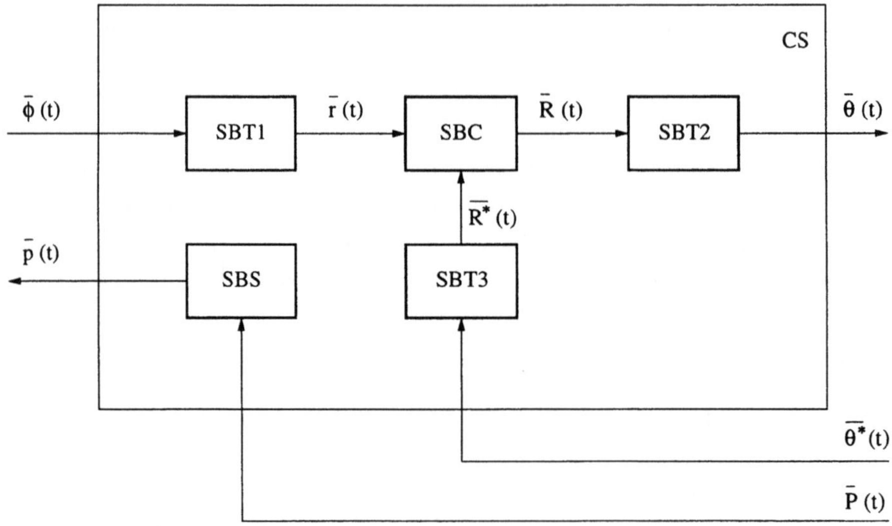

Figure 2.10. Structure of a manipulator control system. Designations: CS is a control system; SBC is a sub-block of the actual remote control; SBT1 is a sub-block that solves a direct ID kinematic problem; SBT2 is a sub-block that solves an inverse robot kinematic problem; SBT3 is a sub-block that solves a direct robot kinematic problem; SBS is a sub-block SBS that transforms force sensing signals from the robot EO; $\vec{\phi}(t)$ is a vector of ID joint coordinates; $\vec{r}(t)$ is a vector of components of ID Cartesian coordinates; $\vec{R}(t)$ is a vector of manipulator gripper Cartesian coordinates; $\vec{\Theta}(t)$ is a vector of EO joint coordinates; $\vec{\Theta}^*(t)$ is an actual vector of manipulator joint coordinates; $\vec{R}^*(t)$ is a vector of manipulator gripper Cartesian coordinates; $\vec{P}(t)$ are force sensing signals entering from the robot EO; and $\vec{p}(t)$ is a vector of signals that enter inputs of a sensing unit.

place in robotics. Therefore, different types of these systems must be described. However, before this, we would like to illuminate another important side of this system application; that is, to say, the problem of their quality.

References

Anderson, D.E. (1989) Telerobotic application to EVA, *Proc. of 2nd Eur. in-Orbit Oper. Technol. Symp.*, Toulouse, 12-14 Sept., Paris, Noordwijk, pp. 115-121.

Andreenko, S.N., Voroshilov, M.S., and Petrov, B.A. (1975) *Design of Manipulator Drives*. Mashinostroenie, Leningrad (in Russian).

Asada, H. and Slotine, J.J. (1986) *Robot Analysis and Control*. John Wiley & Sons, New York.

Backhouse, C.J., Price, R.A., and Moore, P.R. (1994) Crane automation, *Industrial Robot*, **Vol. 21 no. 4**, pp. 22-25.

Bejczy, A.K. (1994) Toward advanced teleoperation in space, in S.B. Skaar and C.F. Ruoff (eds.), *Teleoperation and Robotics in Space*. American Institute of Aerodynamics and Astrodynamics, Progress in Aeronautics and Astronautics Series, AIAA, Washington,

DC, pp. 107-138.

Brimley, W., Brown, D., and Cox, B. (1994) Overview of international robot design for Spase Station Freedom, in S.B. Skaar and C.F. Ruoff (eds.), *Teleoperation and Robotics in Space.* American Institute of Aerodynamics and Astrodynamics, Progress in Aeronautics and Astronautics Series, AIAA, Washington, DC, pp. 411-441.

Burdea, G. and Coiffet, P. (1994) *Virtual Reality Technology.* John Wiley & Sons, New York.

Cardaun, U. (1990) Erfahrung vorausgesetzt: Off-line-Programmierung von Lackierrobotern, *Roboter*, **no. 5**, pp. 16-18 (in German).

Chen, L.-Y., Fujimoto, H., Yagihashi, M., and Kakita, N. (1995) New robot teaching system with operating stick using the Virtual Reality technology, *Proc. of 7th Intern. Conf. on Advanced Robotics*, Spain, Sept. 20-22, **Vol. II**, pp. 607-612.

Craig, J. (1986) *Introduction to Robotics: Mechanics and Control.* Addison-Wesley, Reading, MA.

Fu, K.S., Gonzalez, R.C., and Lee, C.S.G. (1987) *Robotics: Control, Sensing, Vision and Computer.* McGraw-Hill, New York.

Goldenberg, A.A., Wiercienski, J., Kuzan, P., Szymczyk, C., Fenton, R.G., and Shaver, B. (1995) A remote manipulator for forestry operation, *IEEE Trans. on Robotics and Automation*, **Vol. 11 no. 2**, pp. 185-197.

Green, P.S., Hill, J.W., Jensen, J.F., and Shah, A. (1995) Telepresence surgery, *IEEE Engineering in Medicine and Biology Magazine*, **Vol. 14 no. 3**, pp. 324-329.

Herndon, J.N. and Hamel, W.R. (1987) Telerobotic technology for nuclear and space applications, *AIAA Pap.*, **no. 1690**, pp. 1-12.

Hirchfeld, R.A., Aghazadeh, F., and Chapleski, R.C. (1993) Survey of robot safety in industry, *The Intern. J. of Human Factors in Manufacturing*, **Vol. 3 no. 4**, pp. 369-379.

Judd, R.P. and Knasinski, A.B. (1987) A technique to calibrate industrial robots with experimental verification, *Proc. of IEE Intern. Conf. Rob. and Autom.*, Raleigh, NC, March 31-Apr. 3, **Vol. 1**. Washington, DC, pp. 351-357.

Knape, B.P. and Berger, A. (1991) Development of a remote tank inspection (RTI) robotic system, in M. Jamshidi and P.J. Eicker (eds.), *Robotics and Remote Systems. Proc. of Fourth ANS Topical Meeting on Robotics and Remote Systems*, pp.471-481.

Langner, K. (1986) Nur Schweissrobotern mit wirtschaftlicher Programmirung gehort die Zukunft, *Fachber. Metallbearb.*, **Vol. 63 no. 7-8**, pp. 346-347 (in German).

Morley, E.C. and Syan, C.S. (1995) Teach pendants: How are they for you? *Industrial Robot*, **Vol. 22 no. 4**, pp. 18-22.

O'Hara, J.M. and Olsen, R.E. (1988) Control device effects on telerobotic manipulator operations, *Robotics and Autonomous Systems*, **Vol. 4 no. 1**, pp. 5-18.

Ollend, S. (1989) Robotics program at the NASA Goddard Space Flight Center, *Proc. of 2nd Eur. in-Orbit Oper. Technol. Symp.*, Toulouse, 12-14 Sept., 1989. Paris, Noordwijk, pp. 13-21.

Roman, H.T. (1989) Robots shine in electric industry, *Robotics Today*, **Vol. 2 no. 3**, pp. 2-4.

Ruoff, C.F. (1994) Overview of space telerobotics, in S.B. Skaar and C.F. Ruoff (eds.), *Teleoperation and Robotics in Space.* American Institute of Aerodynamics and Astrodynamics, Progress in Aeronautics and Astronautics Series, AIAA, Washington, DC, pp. 3-21.

Schraft, R.D. and Wanner, M.C. (1993) The cleaning robot "Skywash," *Industrial Robot*, **Vol. 20 no. 6**, pp. 21-24.

Schreiber, R.R. (1984) How to teach a robot, *Robotics Today*, **Vol. 6 no. 3**, pp.51-56.

Schwartz, G. (1985) Von der Online- zur Offline-Programmierung von Industrierobotern, *Automatisierungstechn. Prax.*, **Vol. 27 no. 8**, pp. 392-393 (in German).

Springer, S. L. and Gadh, R. (1996) State-of-the-art virtual reality hardware for computer-aided design, *J. of Intelligent Manufacturing*, **Vol. 7**, pp. 457-465.

Sheridan, T.B. (1992) *Telerobotics, Automation, and Human Supervisory Control.* The

MIT Press, Cambridge, MA.
Thring, M.W. (1983) *Robots and Telechirs: Manipulators with Memory; Remote Manipulators; Machine Limbs for the Handicapped.* Ellis Horwood Ltd, Chichester.
Vertut, J. and Coiffet, P. (1984) *Teleoperation and Robotics: Evolution and Development*, Vol. 3A. Prentice-Hall, Englewood Cliffs, NJ.
Vertut, J. and Coiffet, P. (1985) *Teleoperation and Robotics: Applications and technology*, Vol. 3B. Prentice-Hall, Englewood Cliffs, NJ.
Whitney, D.E. (1972) The mathematics of coordinate control of prosthetic arms and manipulators, *Trans. ASME, Ser.G., J. Dynam. Sys. Meas. Control*, **Vol. 94 no. 4**, pp. 303-309.
Ziegert, T. and Datseris, P. (1989) Basic considerations for robot calibration, *Intern. J. Rob. And Autom.*, **Vol.4 no.3**, pp. 158-166.

CHAPTER 3

QUALITY ANALYSIS AND DESIGN OF THE REMOTE MANIPULATION SYSTEMS

3.1. Approaches to the Remote Control Quality Problem

Various approaches to quality evaluation and improvement of operator-remote manipulator systems are now in use in robotics which are applied to the different kinds of problems to be solved. These approaches are used to estimate pre-project solutions that may be used in further robot design. This stage is usually performed by applying modeling and simulation methods, while methods of semi-natural simulation are usually used when a real HO controls a robot model. Complete simulation may also be applied when both models - the HO and robot - are in use; however, the adequacy of this approach, when using so many complicated controlled objects such as robots, must always be verified.

When simulating remote manipulation systems, telerobot models of different complexity may be used, from the most simple to the most complicated. The problem of choosing the optimal complexity of the model always arises when simulation methods are applied. As this takes place, use of the simplest model allows us to better study the essence of physical processes proceeding in the system and permits a better understanding of general conformities to natural laws of a system behavior. This was the method applied in our studies, described below.

Another approach is related to the use of a very detailed robot model (see, e.g., (Medvedev *et al.*, 1978)), which aims at simulating a concrete manipulation system, determining its characteristic, and using them to design an actual system. While this approach is very profitable, it is also too labour consuming. The problem is that, by using both the simple and complicated models, the final evaluation of a developed telerobot has nevertheless to be performed in real conditions of its testing and exploitation.

The second approach towards the quality evaluation of remote manipulation system relates to the study of actual telerobots. Modern, manually controlled manipulators are characterized by a great number of DOFs, complexity of fulfilled tasks, and the continuos intervention of a HO, whose task consists of building the control programs and realizing them. To estimate the quality of these highly functional and peculiar MMSs, as well as that of other manual control systems, the subjective qualitative opinions of operators (such as "easily," "complicated," "hard") were used as a basis for evaluation. This period, however, ended in the early 1970s and gradually investigators and designers began to use methods that enabled them to form objective quantitative estimates of remote control quality.

Note that the methodology for remote manipulation system testing is roughly the same as that for IR testing; therefore, almost all known results developed to test IRs may be used to test teleoperators. A peculiarity of the latter problem is, we think, that telerobots perform more various operations, and therefore means for their testing can be more refined in order to check the teleoperator more comprehensively.

3.2. Experimental Determination of Telecontrol Quality

3.2.1. KINDS OF MANIPULATION TASKS

Since methods for theoretical design of telerobotic systems are not yet perfect, experimental investigation remains a principal means of determining system quality. Telerobot quality can be determined through the manipulation operations that the robot typically performs. Unit operations selected for this purpose may include (Slutski, 1980):

O_1: to grasp (to release) an object to be manipulated;

O_2: to transfer an object (or a gripper without the object) on an arbitrary trajectory;

O_3: to transfer an object (or a gripper without the object) on a prescribed trajectory;

O_4: to change the object angle coordinates; that is, to turn it (e.g., to screw a nut on a screw).

This set of primitive tasks is almost the same as that proposed by Hill (1979), which appears in materials presented by Book (1985). Note that Hill's tasks of inserting and disengaging appear to be combinations of our unit tasks O_3 and O_4; that is, each operation of, for example, inserting may consist of a sequence of unit tasks of a specific turn of an object (O_4), followed by transfer of the object on a prescribed trajectory (in this case, a straight line - O_3).

Based on these unit tasks, the quality of a manipulation system can be determined by using corresponding quality criteria.

3.2.2. QUALITY CRITERIA

General Approach

To determine the quality of telerobotic systems, we developed a method (Kobrinskii *et al.*, 1971; Kobrinskii *et al.*, 1974; Slutski, 1972) based on the idea that the quality of fulfillment of a great number of technological operations associated with the use of manual labor is characterized by two factors: accuracy, and speed, all other factors being equal. In fact, a vector of criteria can be organized that includes, not only these two criteria for a task completion (time $I_1 = T$, and an accuracy estimate $I_2 = \Phi$), but also other system indices, such as reliability, stability (Slutski, 1980). As a result, a vector of the quality criteria of a given system can be formed as $\vec{I} = (I_1, I_2, ..., I_p)$. A feature of this vector is that its last $p-2$ components are, as a rule, functions of first two components

$$I_{2+j} = I_{2+j}(I_1, I_2), \ (j = 1...p-2).$$

In fact, the indices of accuracy and fast-operation may serve as indirect criteria of system reliability. For example, their abrupt increase while the system is working would testify that system reliability was lost. Similarly, changes in the system's dynamic errors enable one to judge the system's stability. This reasoning allows us to reduce the dimensions of vector $\vec{I}$ and to limit ourselves to only the two components, time and accuracy.

In accordance with this premise, the criteria were chosen, including accuracy and time indices of MMS operation. An experimental study was then performed, the results of which are discussed below.

Accuracy Characteristics

A telerobot's accuracy may be evaluated using control theory, which states that there are two principal kinds of control problems: terminal, and path. This division is suitable for such controlled objects as robots. It can be seen that the elementary robotic operations O_1,...,O_4 belong to these two classes of control problems. Thus, operations O_1, O_2, and O_4 are examples of terminal control problems and operation O_3 belongs to the class of path control problems. Correspondingly, errors raised are also related to two different types, which Liegeois (1985) classified as static, and dynamic.

Without going into details, for terminal problems (what are called targeting problems in telerobotics, or point-to-point control problems for IRs), errors are determined with the help of two values: accuracy, and repeatability. The Robotic Industries Association defines these terms as:

- **accuracy** is the degree to which actual position corresponds to a commanded (desired) position;

- **repeatability** is the closeness of agreement of repeated position movements, under the same conditions, to the same location.

These error formulations are discussed by Liegeois (1985), who presents them in detail. We would like to note only that the accuracy of the targeting task is characterized by mean values of positioning errors raised and measured during multi-time performance of this task by a manipulation system.

For path, or (in robotics) contouring, control problems, accuracy index may be calculated by considering the fact, while tracing a reference trajectory, a certain surface is formed that is located between the reference trajectory and actual trajectory realized by a telerobot endpoint (Slutski, 1972). This surface area characterizes tracing accuracy.

In fact, when performing some identical cycles of path operations, all spatial trajectories of robot EE are inside of a certain tube, and the characteristics of this tube have to be measured to estimate robot accuracy. However, this solution is too complicated from a technical viewpoint. Therefore, a planar reference task can be considered. In this case, the solutions to the technical problems of measurements are much simpler.

To trace a certain plane reference trajectory, a picture may be obtained, as shown in Figure 3.1. The area between the two curves may then be calculated, for example, with the help of a contour integral of the second kind. For the two-dimensional case this integral, which is a function of $f(x, y)$ for a continuous or piecewise continuous curve AB (Fig. 3.1), can be written as

$$\int_{AB} f(x,y)dl = \lim_{\Delta l_{imax} \to 0} \sum_{i=1}^{n-1} f(x_i, y_i)\Delta l_i. \tag{3.1}$$

Here $f(x, y)$ is the distance from the actual trajectory $AA_i'B$ to the reference trajectory AA_iB (a line A_iA_i' is perpendicular to the curve AA_iB), and Δl_i is the length of the arc $A_{i-1}A_i$.

Depending on the kind of reference curve setting, integral (3.1) can be reduced to a definite integral. For example, a calculation to evaluate accuracy is essentially simplified for reference trajectories when polar coordinates are used. The reference trajectory is then set by equation $\rho_r = \rho_r(\phi)(\phi_0 \leq \phi \leq \phi_1)$, and (3.1) is rewritten as

$$\int_{AB} f(x,y)dl = \int_{\phi_0}^{\phi_1} f[\rho_r(\phi), \phi]\sqrt{\rho_r^2 + \rho_r'^2}d\phi. \tag{3.2}$$

As is known in principle, not only the integral index, but the value of the maximum deflection from the reference trajectory of a path realized by the HO may also be used as an accuracy characteristic of such a tracing.

As for repeatability, it is usually expressed for both targeting and tracing manipulation problems as a value equal to three times the standard deviation of the task errors (Liegeois, 1985). Concerning all quantitative calculations of accuracy and repeatability indices, we can refer to the International Standard (ISO, 1990), where they are presented in detail.

Note that, derivatives of all these kinds of errors relative to time may also be calculated and used if needed. These are useful to estimate dynamic properties of a robot.

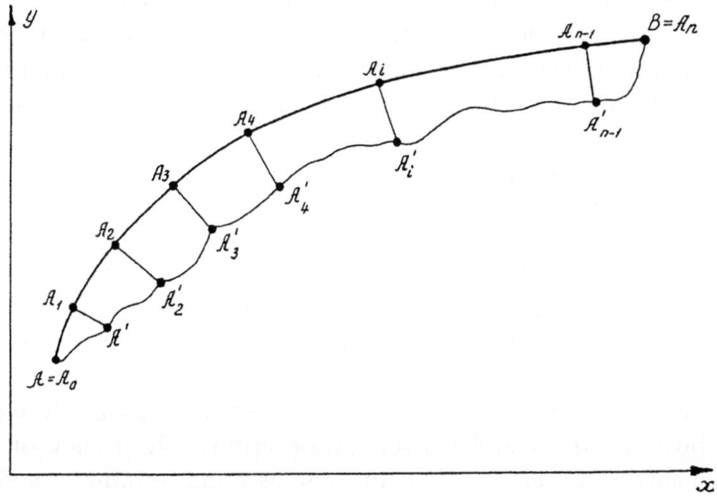

Figure 3.1. Scheme of accuracy criterion calculation for a path control problem. Designations: AA_iB is a reference trajectory and $AA_i'B$ is an actual trajectory.

Complex Criterion

A problem of multicriteria evaluation with contradictory criteria was formulated above when two system criteria of task completion time $I_1 = T$ and accuracy estimate $I_2 = \Phi$ were introduced. The most practical solution of these problems usually leads to a convolution of this vector's components in a general scalar index, which may be further used both to estimate system quality and to synthesize it.

The basis of such a convolution may be an index (further on it will be called complex criteria), according to (Kobrinskii *et al.*, 1971; Kobrinskii *et al.*, 1974), the product of two values, one of which, $F(T)$, is a function of time T of operation fulfillment; and the second, $\Phi(\epsilon)$ is the accuracy characteristic (ϵ is the current error):

$$J = F(T)\Phi(\epsilon). \tag{3.3}$$

As for $F(T)$, its form depends on the weight given to this component of the complex criterion (3.3).

Another form of the complex criterion used was the following:

$$P = \lambda_1 \frac{F(T)}{F_0} + \lambda_2 \frac{\Phi(\epsilon)}{\Phi_0}, \; \lambda_1 + \lambda_2 = 1, \tag{3.4}$$

where λ_1 and λ_2 were the weight coefficients and F_0^{-1} and Φ_0^{-1} were normalizing multipliers (Slutski, 1972).

Concerning the application of two forms of the complex criteria, it is necessary to note that the first criterion (3.3) has a theoretical shortcoming: it gives a displaced estimate in certain cases. According to this criterion, the best evaluations may be obtained when extreme strategies are in use; that is, when a task is performed either with the minimum error and extended period, or with minimum time and larger error. Criterion (3.4) does not have this shortcoming. There is, however, another difficulty in using it; the choice of normalizing multipliers presents a great practical problem. This problem concerns not only criterion (3.4), but also other indices based on measurements of "distance" from a current point in a criteria space to a certain "ideal" point. Thus, a decision was made to choose for our investigation criterion (3.3), particularly since the experimental results indicated that the HO does not choose extreme strategies and always tries to optimize both accuracy and fast-operation criteria. Efficiency of criterion (3.3) when solving problems of quality analysis and parameter synthesis of remote manipulation systems using the method of the semi-natural simulation is described below.

Moreover, it was previously shown by us (Kojinskaya and Slutski, 1973) that there is a certain tie between these two kinds of criteria and, more generally, between a linear convolution of a general type (3.4) and some another concrete type of the convolution of the criteria vector into scalar index. It was demonstrated that, in this case, results of optimization by criterion (3.4) with a determined set of weight coefficients will be identical with results of optimization by another type of complex criterion (e.g., (3.3)).

3.2.3. QUALITY EVALUATION PROCEDURES: SIMULATION

Experimental Procedure

To simulate remote control processes based on the task of tracing a plane reference curve, the accuracy estimate (3.2) may be used; however, in practice, it is more convenient to calculate an integral quadratic functional. The estimate of the accuracy characteristic is then calculated by the formula

$$\Phi = \int_{\phi_0}^{\phi_1} (\rho - \rho_r)^2 \sqrt{\rho^2 - \rho_r'^2}\, d\phi, \tag{3.5}$$

where

ρ_r is the module of the reference trajectory radius vector;

ρ is the module of the radius vector of the trajectory realized by the MMS;

ϕ is the polar angle, $\rho_r' = \frac{d\rho_r}{d\phi}$.

The estimate calculation and the manipulator two-coordinate model drives were simulated on an analog computer with the possibility of both

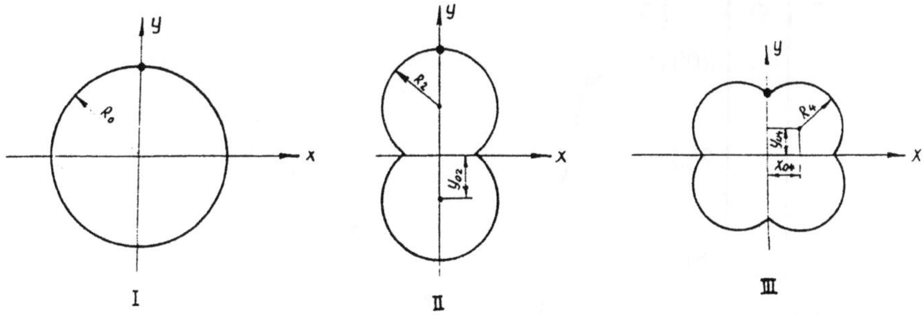

Figure 3.2. Reference trajectories set for a tracing problem. From (Kobrinskii *et al.*, 1974); ©CISM.

position and rate controls of an output member of the x-y recorder, which was used as a model for manipulator EO. An ID - an x-y plane carriage with a handle was installed at the analog computer input. Each coordinate of the handle was furnished with a motion transducer.

HOs were asked to manipulate the control handle and, using the output member of the plane manipulator model, to trace a reference trajectory with minimum errors in a minimum time.

Three trajectories of equal length were taken as references (Fig. 3.2). The equations of radius vectors of these trajectories are in the form:

$$\rho_{rI} = R_0,$$
$$\rho_{rII} = y_{02}|\sin\phi| + \sqrt{R_2^2 - y_{02}^2\cos^2\phi}, \qquad (3.6)$$
$$\rho_{rIII} = 2R_4 \sin 45^o(|\sin\phi| + |\cos\phi|).$$

The choice of trajectories I, II, III was made for the following reasons: a) the comparative simplicity of setting-up the given configurations on the analog computer; and b) the possibility of the relatively simple estimation of the task; that is, each curve complexity.

According to the method suggested by Ganzen and Granovskaya (1965), an estimation of complexity of a certain configuration is based on three characteristics: the number of changes of curvature sign on the configuration (U), the number of curvilinear sections of the configuration (V), and the sum of absolute increment of the inclination angle of the tangent (W). Using these characteristics, we introduced a generalized estimate of configuration complexity: $Z = U + V + \frac{W}{\pi}$. The following estimates of the geometrical complexity were then obtained for reference trajectories I, II, III, respectively: $Z_1 = 3$; $Z_2 = 9.5$; and $Z_3 = 17$.

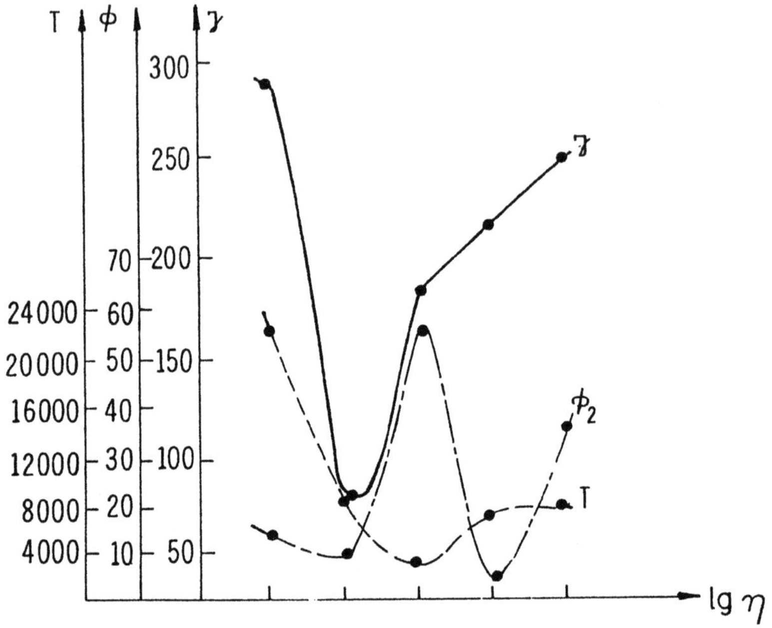

Figure 3.3. Task completion time T, tracing error Φ and complex criterion J (3.3) vs. the gain η of the rate system (all quality indexes are measured in relative units). From (Kobrinskii *et al.*, 1974); ©CISM.

Experimental Results

The initial stage of the operator-manipulator system study was that of operator (5 males) training. This stage was continued for approximately 25-30 cycles of tracing.

After this, experiments were performed to study the relationship of the values of the complex criterion and its components in changing the gain of the x-y manipulator (in terms of the gain in the rate system, the relation of the motion speed of the output member to the displacement value of the manipulator ID is understood; in the case of position control, the gain is the scale ratio).

Results of the statistical processing of the experimental data revealed that, in the rate control method with the large gain, it is natural for the HO to decrease the tracing speed of the reference trajectory. Therefore, the dependence of task fulfillment time on the gain is of an extreme nature (Fig. 3.3). The abscissa of this graph is the logarithms of the following values of gain η in 1/sec: 0.0675; 0.213; 0.675; 2.134; and 6.75. Each point on this typical graph is the averaging result of one of the skilled operators over ten experimental values. The factor of the tracing speed decrease also

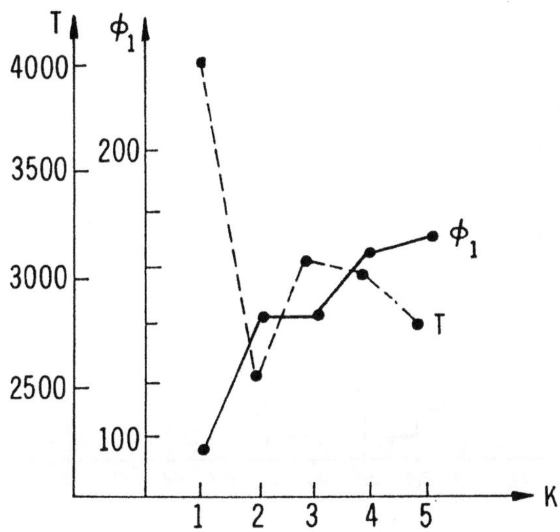

Figure 3.4. Task completion time T and tracing error Φ_1 (measured in relative units) vs. the gain k of the position system. From (Kobrinskii *et al.*, 1974); ©CISM.

has an influence on the dependence of the tracing accuracy (3.5) on the gain, which has local extremums (Fig. 3.3). In this case, the dependence of complex criterion (3.3) (when $F(T) = T$) on the gain is also of a similar nature.

In studies of manipulator position control, some changes have been introduced into the criterion calculation scheme. The quality of task fulfillment was estimated by the simplified (as compared to (3.5)) integral criterion

$$\Phi_1 = \int_0^T (\rho - \rho_r)^2 d\tau.$$

It turns out that, for a position control system, the first approximations of the integral estimate Φ_1 and tracing time T have, respectively, direct and inverse (Fig. 3.4) dependences on the model gain.

A comparison of task fulfillment results by trained operators, made with the help of the complex criterion, permits us to estimate the skill of each operator. Figure 3.5 shows the histograms obtained by processing the results of the reference trajectory tracing by three operators, 50 times each.

The study of the influence of task complexity - that is, geometrical complexity of the reference trajectory - on tracing quality has indicated increase in tracing time as the curve complexity increases. At the same time, one should note the fact that there is no distinct tendency in the

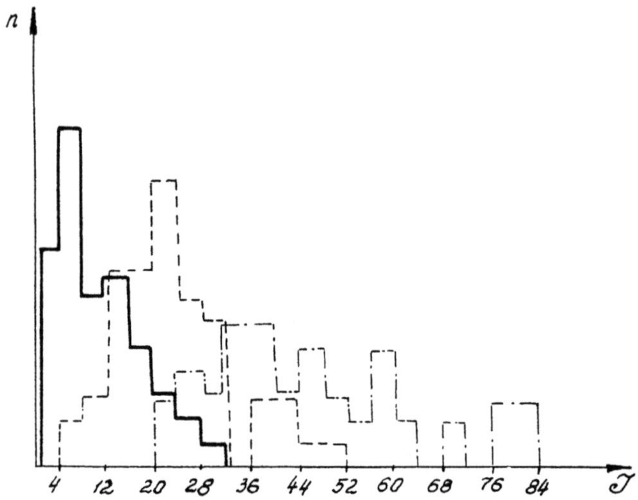

Figure 3.5. The histogram of the error distribution for the reference trajectory tracing, J is complex criterion (3.3) (dashdot line - operator #1; dashed line - operator #2; solid line - operator #3). From (Kobrinskii *et al.*, 1974); ©CISM.

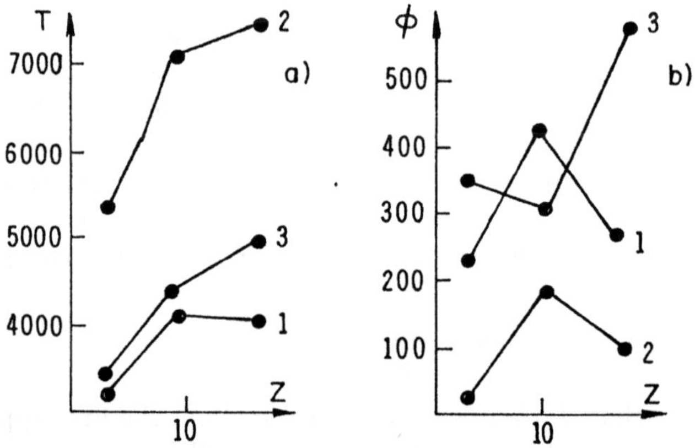

Figure 3.6. Task completion time T and tracing error Φ (in relative units) vs. the geometrical complexity Z of the reference trajectories. From (Kobrinskii *et al.*, 1974); ©CISM.

changing tracing accuracy to depend on configuration complexity; e.g., the graphs of Figure 3.6 show both a decrease of accuracy with a complication of trajectory for one operator, and an improvement of tracing results, under the same conditions, for another operator. It can be concluded that an operator tries to keep the tracing error on a certain constant (evidently nominal for him) level at the expense of the time required for this purpose.

Thus, the accuracy estimate is a major factor in choosing control strategies for each operator in the tracing task. The same conclusion was also made from a learning process study (Kobrinskii *et al.*, 1974).

3.2.4. TESTING IMPLEMENTATION FOR ACTING ROBOTIC SYSTEMS

General Basis of Stand Testing

Among different approaches to checking acting robotic systems, one is now under development that is based on telerobot testing in real environments. For this, it is necessary to complete the robotic system with actual technological equipment, to create an industrial locality rigged with real details and auxiliary devices and to reproduce the actual technological process during the experiment. The use of such localities permits one to decide a large variety of robotic system work questions. This approach has been used in a number of investigations of remote manipulation systems (e.g., (Wilt *et al.*, 1977)). It should be particularly emphasized that this method gives quite reliable and practical results, but it is very expensive and does not usually allow one to precisely measure the accuracy and productivity characteristics of a manipulation system.

Another approach to manipulation system checking is solving test problems. For remote manipulation systems, these investigations were initialized and described in (Kugath and Wilt, 1974), among others. For IRs, similar studies have been very actively conducted for a long time by Prof. Warnecke and his colleagues at the Institute for Production and Automation in Stuttgart (see, e.g., (Warnecke and Schiele, 1985)), as well as in other countries (e.g., (Ranky and Ho, 1985)). Some of these testing schemes were designed in the Soviet Union beginning from 1977 when the first invention was registered (Slutski, 1977). The reason for the intensive development of this method is that it is faster and cheaper compared to other methods, and it gives precise quantitative estimates of telerobot (or IR) quality, as well as allowing rapid rearrangement of robot locality and organization of new operation executions.

Note that most overall investigations of telerobots, that do important tasks (e.g., space applications), include elements of both actual technological tasks and testing operations (Hannaford *et al.*, 1991). In these experiments, four different tasks were used to check the developed teleoperator quality:

Task 1 - Velcro Attachment

Task 2 - Peg-in-Hole Matrix

Task 3 - Electrical Connectors

Task 4 - Bayonet Connector

It may be seen that Tasks 1, 3, and 4 are real manipulation tasks which may be performed in the operational environment of any actual telerobot. At the same time, Task 2 is essentially a testing task and was carried out by inserting a standard peg into nine holes with different diameters. This checking was done to study a force-reflecting teleoperator; therefore, measurements were performed based on the data of a six-axis force-torque sensor which was mounted in series with a gripping mechanism. However, the same index of completion time was used as a principal characteristic of system quality. An additional quantitative index - the sum of squared forces (SOSF) - was used that reflects a specific detail of the studied system. The SOSF was computed by taking a nondecreasing sum of the square of the force or torque value:

$$SOSF = \sum_{i=1}^{N} f_i^2 dt,$$

where N is a number of data samples (task time over dt), f_i is the ith sample of force or torque, and dt is the sampling interval. SOSF was accumulated separately for each force and torque axis.

In this study, very productive methods of task segmentation analysis were performed. This kind of analysis was used to compare performance measures among different segments of the same task; e.g., a comparison was carried out for the completion time and SOSF for peg insertion with peg extraction.

From the results of this research, it would appear that test investigation data appear to give a sharper picture of differences between different types of systems. The reason may be that it is difficult to precisely measure task completion time for such operations as, for instance, a velcro attachment.

The general principle of testing manipulation systems (including IRs) with test stands requires us to introduce a quality evaluation criteria system and to model a set of elementary manipulation operations (Slutski, 1980).

For the quality fulfillment analysis of manipulation problems, there are five possible combinations of accuracy and operation time indices for a telerobotic system that coincide with the versions of task formulation:

C_1: to fulfill the task in a minimum time with predetermined accuracy;

C_2: to fulfill the task with minimum errors in a predetermined time;

C_3: to fulfill the task with minimum errors in a limited time;

C_4: to fulfill the task with limited errors in a minimum time;

C_5: to fulfill the task minimizing both criteria.

The tasks $C_3(\Phi \rightarrow min, T < T_L = const)$ and $C_4(T \rightarrow min, \Phi < \Phi_L = const)$ require operation fulfillment with the availability of one criterion and one limitation. The tasks $C_1(T \rightarrow min, \Phi = const)$ and $C_2(\Phi \rightarrow min, T =$

const) are generally reduced to one criterion evaluation. The formulation of $C_5(\Phi \rightarrow min, T \rightarrow min)$ presents the problem of a multicriteria evaluation. However, the most practical idea for setups of telerobot testing that of fixing one of these criteria and measuring another, which allows one to obtain rather determined results and to use them for comparing different types of telerobotic systems.

Each of above formulations of the tasks to be executed may be used in the quality evaluation process to carry out operations O_1,...,O_4.

The test experiment organization depends on the combination of the performance evaluation versions (C_1,... ,C_5) and the kind of tasks to be performed (O_1,... ,O_4). Integration of these versions and the elementary manipulation task combinations gives 75 possible setups for the efficiency evaluation. In Table 3.1, the following designations for operation combinations are used:
$T_1 = O_1$; $T_2 = O_2$; $T_3 = O_3$; $T_4 = O_4$;
$T_5 = O_1 \wedge O_3$; $T_6 = O_1 \wedge O_2$; $T_7 = O_1 \wedge O_4$; $T_8 = O_2 \wedge O_3$; $T_9 = O_2 \wedge O_4$; $T_{10} = O_3 \wedge O_4$;
$T_{11} = O_1 \wedge O_2 \wedge O_3$; $T_{12} = O_1 \wedge O_2 \wedge O_4$; $T_{13} = O_1 \wedge O_3 \wedge O_4$; $T_{14} = O_2 \wedge O_3 \wedge O_4$; $T_{15} = O_1 \wedge O_2 \wedge O_3 \wedge O_4$.

The symbol $\wedge$ designates a logical function "AND." The following designations of sources and setups described below are used in Table 3.1: W - (Warnecke and Schiele, 1985), H - (Hannaford *et al.*, 1991), V - Vukobratovic's drink-test, (1976), K - (Kobrinskii *et al.*, 1974), 1 - Setup S-1, 2 - Setup S-2, 3 - Setup S-3, 5 - Setup S-5, 6 - Setup S-6.

TABLE 3.1. Combinations of practical schemes for remote manipulation system testing with certain criterial estimates

Tasks	T_1	T_2	T_3	T_4	T_5	T_6	T_7	T_8	T_9	T_{10}	T_{11}	T_{12}	T_{13}	T_{14}	T_{15}
Estimate															
C_1			6			2	1						V		3
C_2			W												
C_3			W												
C_4			W		5			H	W						
C_5		W	K												

Each item of this table means a certain experimental setup that can be realized both in conditions of an experimental simulation or of real system testing. A number of schemes have been suggested in the scientific literature and patents, but only some of these setups are noted in the table as examples. A presentation of all known setups would make the table too unwieldy. It is obvious for us that the classification in Table 3.1 may be

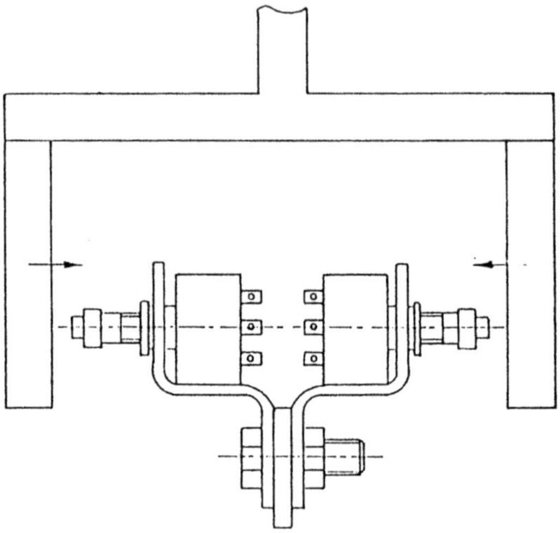

Figure 3.7. Layout of a position sensor to simulate grasping an object to be manipulated.

performed in different ways; however, our principle goal here was to give an example of how such a classification might look. It is obvious that the technical realization of each scheme may be different.

Test Stand Setups

The following setups are examples of the data presented in Table 3.1.

Setup S-1

The following simple pick-and-place task must be performed: an object must be clutched at a certain point in a robot's workspace and be transported to another point, where it must be released. Instead of clutching a real object, the process of grasping can be simulated. For this, two pushbuttons (or limit switches) have to be united; they must also be placed exactly opposite each other; that is, each pushbutton corresponds to each jaw of a robot gripper (Fig. 3.7). This setup obliges the telerobot operator to perform grasping an object with two jaws of a gripper simultaneously, it prevents from pressing only one pushbutton by, for example, the back of the gripper. The same sensor is located in another point of a robot's workspace. The task is to grasp the first sensor; that is, to precisely perform this action, and to move the gripper further on an arbitrary trajectory to another sensor and grasp it. When the sensors are turned on and off, a clock measures the task completion time.

Setup S-2

Based on the previous setup, objects may be placed in a robot workspace that simulate specific industrial restrictions of the workspace. A gripper

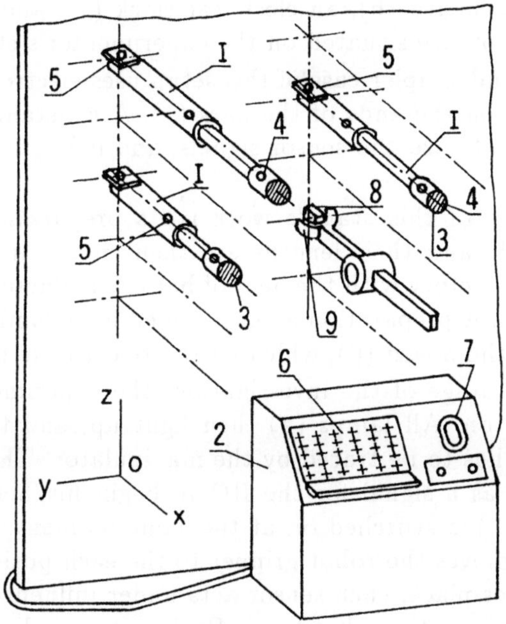

Figure 3.8. Layout of a test stand according to Setup S-4. Designations: 1 are telescopic bars, 2 is an experimenter's station, 3 are sensors, 4 are lamps, 5 are devices for fixing of bar length, 6 is a pin board, 7 is an electrical clock, 8 is a constant magnet, and 9 is a robot gripper.

shift is then set with allowances, which leads to certain demands on shift accuracy.

Setup S-3

This setup differs from the Setup S-1 by a shift of the robot gripper on a certain trajectory, rather then on an arbitrary one. This trajectory can be set by successive location of sensors (Fig. 3.7) in the manipulator workspace; that is, the trajectory is partly prescribed and partly arbitrary (between its node points). The gripper must grasp each of them in turn. The peculiarity of this scheme is that a position sensor axis can be turned according to different axes of the world coordinate system. This does more complicated tasks, requires manipulator orientation motions, and gives a more complete estimate of telerobot characteristics.

Setup S-4

This setup also implements a set of a spatial trajectory, but it uses non-contact position sensors, and, therefore, a grasping operation is not simulated here. A corresponding stand is shown in Figure 3.8. It consists of telescopic bars (1) that are placed in a vertical wall, and an experimenter's station (2). A pin board (6), an electrical clock (7), and buttons to turn the stand on and off are situated on the experimenter's station (2).

A particular and simpler case of this setup uses electromagnetic sensors that are situated on the ends of the bars (1). A constant magnet (8) can then be used to initialize the sensor signals, and it is grasped in the robot gripper (9).

The first stage of this stand's work is its preparation. The required telescopic bars (1) and their lengths are then chosen in order to form a certain spatial trajectory with the help of bar ends. These lengths are fixed by the devices (5). A preparation stage is ended by a commutation of plugs into the holes of the board (6), which correspond to the pulled out bars.

In the second stage of the investigation, the experimenter presses the test start-up button. All lamps (4) then light up, and they delineated a trajectory which has to be traced by the manipulator. The switching on of the lamps serves as a signal for the HO to begin his/her work, since the electronic clock (7) is switched on at the same moment. By performing a tracing, the HO moves the robot gripper to the each position sensor (3) in turn. As this takes place, each sensor acts under influence of the constant magnet (8) clamped in the robot hand. Each corresponding lamp then goes out and the HO can pass to another trajectory point; that is, to another position sensor (3). When all points of the trajectory are passed, all lamps have gone out and the clock (7) stops. Its reading represents a necessary quality estimate.

This setup can be used when testing telerobots without a gripper (for instance, industrial robots for arc welding) or to check a robot's main kinematic chain.

Setup S-5

This setup is intended to study the quality fulfillment of the task of introducing a peg into a hole. The stand (Gamarnik *et al.*, 1989) consists of a bar (2) with a gauging peg (12), which is fixed inside a liner (11) (Fig. 3.9). The latter is primarily placed on a bracket (5) by using a roll cage (3) with a transducer of linear transferences. A touch sensor (13) is located on the peg. The stand also contains a magnet base (6) with attached cup (7). Inside the cup is placed a bushing (8), which is backed with the help of a spring (9). A nut (10) limits the bushing transferences.

While performing testing, a robot grasps the liner (11) and moves it, together with the bar (2), inside the cup (7). When the peg (12) enters the bushing (8), the sensor (13) works on and a measuring device (4) fixes linear transference transducer data by determining the peg's (12) position at the moment. After this, the peg continues its motion together with the the bushing (8) and, in the process, the measuring device (4) continues to fix the linear transference transducer reading.

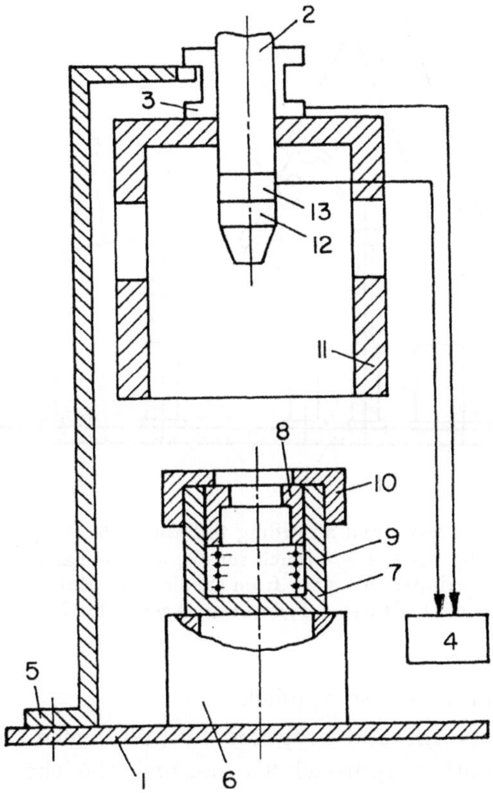

Figure 3.9. Layout of a test stand according to Setup S-5. Designations: 1 is a basis, 2 is a bar, 3 is a roll cage, 4 is a measuring device, 5 is a bracket, 6 is a magnet base, 7 is a cup, 8 is a bushing, 9 is a spring, 10 is a nut, 11 is a liner, 12 is a gauging peg, 13 is a touch sensor. Adopted from (Gamarnik *et al.*, 1989); reproduced here by kind permission of the Russian Federation Agency for Patents and Trade Marks.

The measurement data obtained serves as characteristics of the fulfillment quality of the performed operation. At the same time, a reading value of the linear transference transducer at the moment when the peg (12) touches the bushing (8) characterizes this operation's accuracy fulfillment. Time characteristics can also be determined based on registrated data.

Setup S-6

To organize path control tests, a benchmark-type technology (Warnecke and Schiele, 1985; Wodzinski, 1987) is used as a rule, since it will be the most simple and reliable. Two standard shapes of paths were usually devised - linear and circular - which are typical for robotic applications. These two shapes are sized according to the size of the robot tested. A more gene-

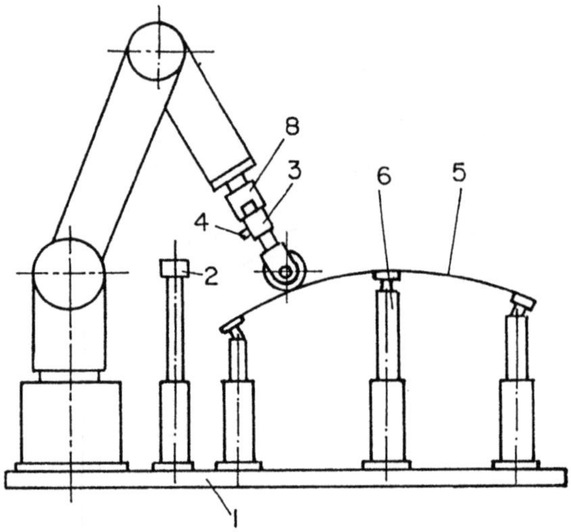

Figure 3.10. Layout of a test stand according to Setup S-6. Designations: 1 is a basis, 2 is a stop, 3 is a measuring head, 4 is a touch sensor, 5 is an elastic tape, 6 are telescopic holders, and 8 is a robot gripper. Adopted from (Sedler and Martynov, 1987); reproduced here by kind permission of the Russian Federation Agency for Patents and Trade Marks.

ral type of a path may be also applied.

In addition to the method proposed by (Warnecke and Schiele, 1985) (see Table 3.1), another approach to measure the characteristics of the contouring task performance may be used (Fig. 3.10). It is based on a tracing trajectory set with the help of elastic tape (5) placed on telescopic holders (6) (Sedler and Martynov, 1987). A touch sensor (4) is attached to the robot gripper (8) as is a measuring head (3) with a spring-backed pulley. At the beginning of a test process, the touch sensor interacts with stops (2) and sends a corresponding signal to a measuring device. This signal may also be taken as the value of the positioning accuracy in this point. Thus, a robot performs a tracing set line and corresponding signals are then produced by a tachogenerator placed in the measuring head and connected with the spring-backed pulley. These signals may be used to estimate both accuracy and time characteristics of the fulfillment of this trajectory task.

Thanks to the use of elastic tape and pull-out holders, different shapes of reference trajectories may be obtained here that ensure more detailed checking of the robotic system.

Setup S-7

To measure both accuracy and fast-operation indices when performing transportation motions, setups with test-elements are applied; that is, a detail of definite form and dimensions that is squeezed in the robot gripper (or is fixed instead of the EE). Some sensors or sensor systems are sure to be placed on the measuring heads at individual points of the robot's accessible region; they are simple and reasonably inexpensive, and they give rather high accuracy, as well as output electrical signals for an automatization of measuring.

When using the test-element, contact measurements are performed as a rule. By and large, a measuring element with a linear transference sensors is employed. These kinds of measurements have frequently been described in the literature (Ranky and Ho, 1985; Warnecke and Schiele, 1985). An example of such a system is shown in Figure 3.11. In this example, a robot holds a test-cube and 3D measurements are performed to obtain indices of repeatability and productivity after computer processing.

Non-contact measurements may also be used in implementing this approach (Ranky and Ho, 1985). These ensure a high precision of measurements due to the use of the more precise sensors.

Our recent elaborations were related to an application of optoelectronic sensors for measuring robot EE positions. Specifically, linear coordinate-sensible transducers were used, which are one-line, silicon photodetectors with a semiconductor structure, which was described in (Ceres *et al.*, 1995), among others.

These elements are very light, precise, and fast to operate, and have resolutions of $(10-20)10^{-6}$ m and a maximum fast-operation of 10^{-4} sec. Therefore, the manipulator arm position is measured very accurately and very quickly.

However, these characteristics imply that the sensors are expensive. An elaboration ceases be cost effective if we place the several measuring blocks at different points and each block contains three or more such sensors. For this reason, we developed the following approach: the measuring head was placed, not at different points of a robot's workspace, but on the robot's hand. This reduced the number of sensors required.

This approach was used to measure three coordinates of an industrial robot gripper position (Fig. 3.12). We present also another example (see Fig. 3.13) where positions of the four coordinates of gripper were measured (Slutski *et al.*, 1989).

Here, the measuring head (1) represents the cylinder on which linear photodetectors (3, 4, 5) are placed. The test-elements (2) in this figure are placed at the different points of the robot's workspace, and there is a disk with a post where diodes (6, 7, 8) with light radiation are placed. During robot testing, the measuring head (1) is brought up by the robot hand to the test-element (2) and the diodes (6, 7, 8) are positioned opposite the corresponding linear photodetectors (3, 4, 5). The distance between them must be less than 10 mm. After the first measurement, the obtained co-

Figure 3.11. Picture of a stand with a test-cube and linear sensors placed on an inverse cube. Dial gauges of Series PM-201 (precision - 10 μm, repeatability - 2 μm) with digital display LCD are from Sylvac SA, Switzerland.

ordinate values are taken as zero values. During the next measurements, the light touch position x relative to the zero position x_0 on the linear photodetector (3) then corresponds to the robot hand displacement along the x axis. The light touch y_1 and y_2 positions on the photodetector (4) give the information about the displacement along the y axis, and angle displacement ϕ, about the x axis. The light touch z position on the linear photodetector (5) corresponds to the displacement along the z axis. This system is intended to test cylindrical coordinate robots. During the testing process, a computer provides the signals for switching on light-diodes in

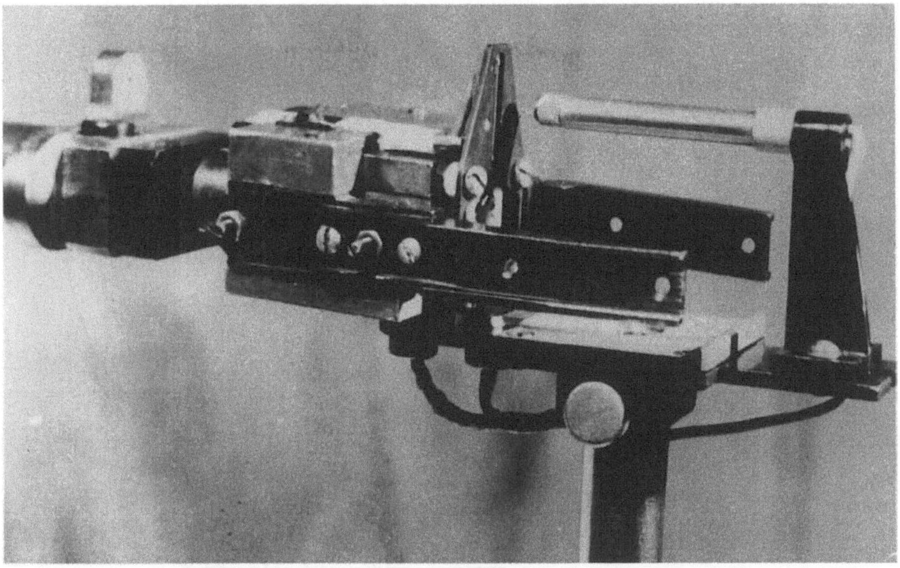

Figure 3.12. Picture of a measuring device with three linear photodetectors and three light-diodes.

turn. The system interrogates the photodetectors; their signals are received through the analog-to-digital converter into the videosignal processing and commutation device and then to a computer. The latter also realizes the transformation operations and statistical processing of the treated measurement information.

In fact, when stand testing is performed, the activity of a corresponding telerobotic system and its environments are simulated. The advantage of using test stands is that all the sensors are usually placed, not on the robot, but on the base, and only very light ones may be located on the robot hand itself. This setup allows faster stand rearrangement leading to faster and cheaper robotic systems testing than can be obtained with other kinds of quality determinations.

3.3. Theoretical Approaches to Study Telerobotic Systems

3.3.1. USING THE MNEMONICABILITY NOTION TO ESTIMATE REMOTE MANIPULATOR QUALITY

The development of theoretical methods of calculation and design of telerobotic systems is very desirable for this field. One such approach is based on the use of a mnemonicability notion, which was introduced for MSSs by Kobrinskii and colleagues (1975). If we designate the vector of the master

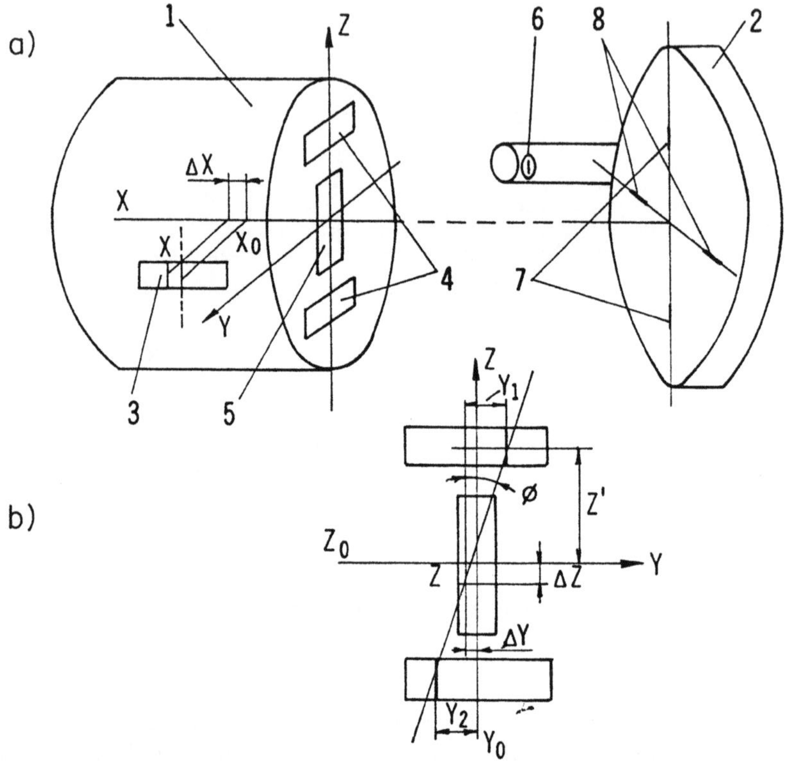

Figure 3.13. Scheme to measure four coordinate accuracy by means of linear photodetectors. Designations: 1 is a measuring head; 2 is a test-element; 3, 4, and 5 are linear photodetectors; 6, 7, and 8 are light-diodes.

endpoint movement by $\Delta\vec{X}$, and the vector of slave gripper point movement by $\Delta\vec{Y}$, then the cosine of the angle between $\Delta\vec{X}$ and $\Delta\vec{Y}$, which is expressed by the formula

$$k_M(\phi, \Delta\vec{Y}) = \frac{(\Delta\vec{X})^T(\Delta\vec{Y})}{|\Delta\vec{X}||\Delta\vec{Y}|} \tag{3.7}$$

for a certain arm configuration ϕ, is named as a local coefficient of mnemonicability (Kobrinskii and Kobrinskii, 1985). This index is introduced for one direction of the vector $\Delta\vec{Y}$. It is apparent that a more complete characterization of mnemonicability for configuration ϕ would be the coefficient

$$k_\phi = \frac{1}{4\pi}\int k(\phi, \beta)d\beta, \tag{3.8}$$

where $d\beta$ is a surface element of the unit sphere. It is also possible to introduce the average value k of the multitude of mnemonicability coeffi-

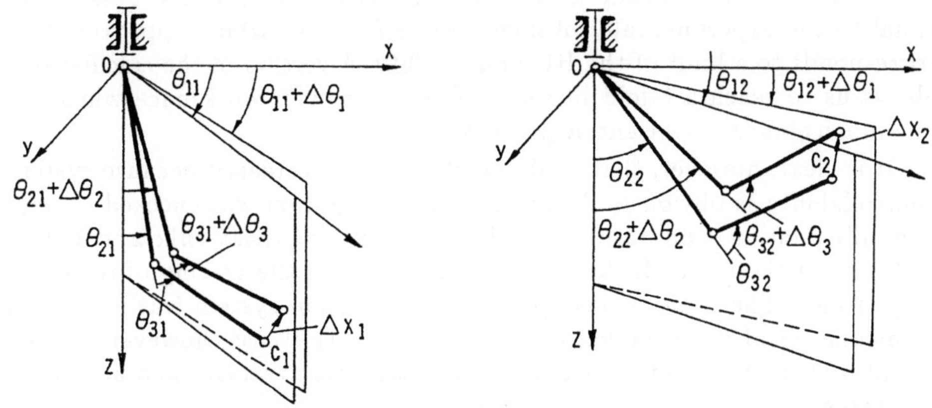

Figure 3.14. Master-slave system with disconcordance on angle Θ_1. From (Kobrinskii and Kobrinskii, 1985); reproduced here by kind permission of Nauka, Moscow.

cients k_ϕ for all possible manipulator configurations. This represents a total estimation of the MSS mnemonicability

$$k = \Lambda^{-1} \int k_\phi d\phi.$$

Here $d\phi$ is an element of volume, and Λ is a complete volume of configuration space. The mnemonicability coefficient gives us a weighted characteristic of a remote control in a manipulation system whose values is changed in the interval $[0, 1]$.

This index was used to estimate the quality of remote control in MSS with disconcordance in positions of master and slave arms. The reason for this change was that conventional MSSs are not always suitable for the HO, because he/she is forced to work in inconvenient positions. Moreover, if there are any restrictions on the HO's working area, a manipulator workspace also turns out to be restricted, which decreases robot's effectivity. Using this disconcordance (Fig. 3.14) allows a considerably increase in the robot's workspace.

These analytical techniques consist of the calculation of mnemonicability coefficients (3.7) and (3.8). The calculations, for example, for the MSS shown in Figure 3.14, give us a value of the coefficient (3.8) that does not depend on a manipulation system configuration and is very simply connected with the variable disconcordance angle Θ_1

$$k_M = k_\phi = \frac{1}{2}[1 + \cos(\delta\Theta_1)].$$

Let a Θ_1 change occur in an interval $|\delta\Theta_1| \leq \delta$, and the δ value be equal to the experimentally obtained value $\delta = 45^o$, which approximately corresponds to a limit of the HO adaptability. A version of the calculations allows us to reach a 3-fold increase of the slave arm workspace when the mnemonicability coefficient $K_M \geq 0.85$.

It is clear, however, that MSS quality is then decreased because system controllability is obviously decreased. This suggestion was checked experimentally, first, in semi-natural simulation conditions (Statsenko and Tyves, 1985) using the methods described above based on the tracing of reference trajectories. The same two-coordinate manipulation system was used with a calculation of the complex criterion (3.3). In this case, however, signals x and y from the input device potentiometers were transformed according to expressions

$$X = k_1(x \cos\gamma + y \sin\gamma), \quad Y = -k_1(x \sin\gamma - y \cos\gamma),$$

where X, Y are signals that enter the x-y recorder from the analog computer; γ is an angle between vectors (x, y) and (X, Y); and k_1 is the gain. The angle γ was changed within limits $-\pi < \gamma < \pi$, and the mnemonicability coefficient was then changed in the range $-1 < k_M < 1$. The investigation was performed for the two first curves of Figure 3.2. For each of these curves, dependences of the complex criterion from the γ value were obtained. Note that this value was changed, as above, in the range of a 15° interval.

For position control, characteristic results were obtained, and a part of these is presented in Figure 3.15, which concern the results of tracing the first reference trajectory - that is, the circle - by four skilled operators. It may be seen that when disconcordances are small ($0^o < \gamma < 45^o$), there is even some reduction of the complex criterion value because of the tracing velocity increase. However, a further increase of the parameter γ value causes a large decrease of tracing quality. When the disconcordance angle is near to 180° and a "mirror control" has place ($k_M = -1$), some improvement of operator work occurs. Also, a worse tracing quality was revealed for negative disconcordance angles ($-60^o < \gamma < 0^o$) than for corresponding positive angles.

Another kind of checking was performed on an actual MSS based on the scheme shown in Figure 3.14 (Kobrinskii *et al.*, 1975). During this experiment, the HOs performed a task to rapidly and precisely move an object from one point to another when some disconcordance values on angle Θ_1 existed. These $\delta\Theta_1 = \Theta_{12} - \Theta_{11}$ values (Fig. 3.14) were set in the range between 0° and 180°. Task completion time was measured as the criterion of operation performance.

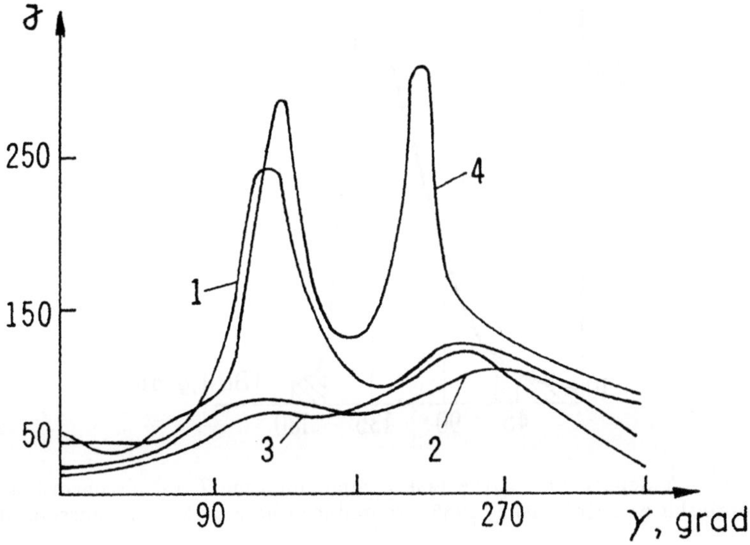

Figure 3.15. Dependences of the complex criterion J vs. disconcordance value γ for four operators. From (Statsenko and Tyves, 1985); reproduced here by kind permission of Nauka, Moscow.

Results of these experiments match well with results of previously described simulation research. They indicated that task completion time was not essentially increased when disconcordance values were approximately less than $45 - 60^o$ (Fig. 3.16). While increasing these values further, the task completion time grew considerably, and for the disconcordance angles $100 - 150^o$, it was increased 15-25 times. While a disconcordance in positions of master and slave arms was mirror-like (near 180°), the HO demonstrated very high accommodation abilities and almost obtained a maximum level of quality of the work.

From these research results, it follows that the HO is able to adapt to the work with a lower level of mnemonicability, and this permits an opportunity to improve telecontrol quality. This low limit of the mnemonicability coefficient was established as $k_M \approx 0.7$ (Statsenko and Tyves, 1985), with the number 0.7 being a border that approximately divides workable and unworkable remote systems.

These results were used to develop remote control systems with a disconcordance of the kinematically similar master and slave that occur when such a system works, in turn, in rate and position control regimes (Statsenko and Tyves, 1985).

We have to remember that the mnemonicability coefficient is, in fact, an

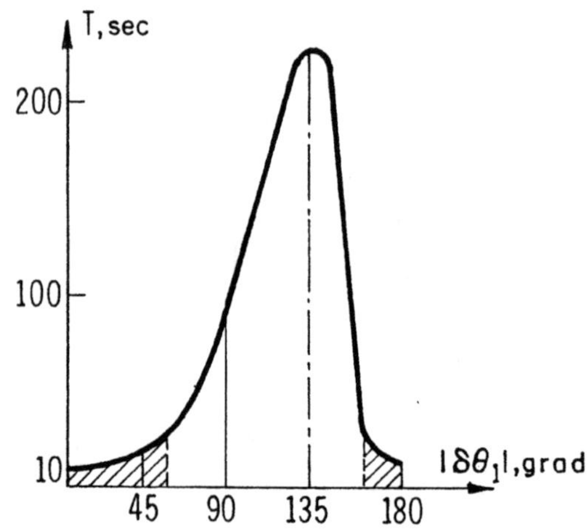

Figure 3.16. A dependence of the task completion time T vs. disconcordance value. From (Kobrinskii and Kobrinskii, 1985); reproduced here by kind permission of Nauka, Moscow.

intermediate index, and it is sometimes difficult to evaluate system quality just by knowing this index value; only if this index is equal to the unit is there maximum system quality. When this index is equal to zero, system quality is at its worst. In all other cases, it is difficult to directly connect the mnemonicability coefficient value with values of telerobot accuracy and rate indices. Therefore, direct resultant estimates of the latter criteria are preferable. However, mnemonicability notion has another positive feature, since it can be used to design remote control systems.

3.3.2. APPROACHES BASED ON FITTS'S LAW

There are presently three approaches to solve problems of remote control system design:
1. based on the use of the HO dynamic model;
2. based on experimental investigations directed toward improving a machine (or its model) during its joint work with the HO;
3. which uses resultant indices such as Fitts's law and its versions.

These approaches are now under development, but it seems to us that the first is currently less in the use in telerobotics. This method deals with dynamic interactions in MMSs (Sheridan and Ferrell, 1974); however, it cannot usually be successfully applied to design control systems for objects with such complicated dynamics as telerobots. Moreover, as a rule, these models do not consider HO adaptation with reference to object control dynamics (Elkind, 1964), although, some investigations have indicated the

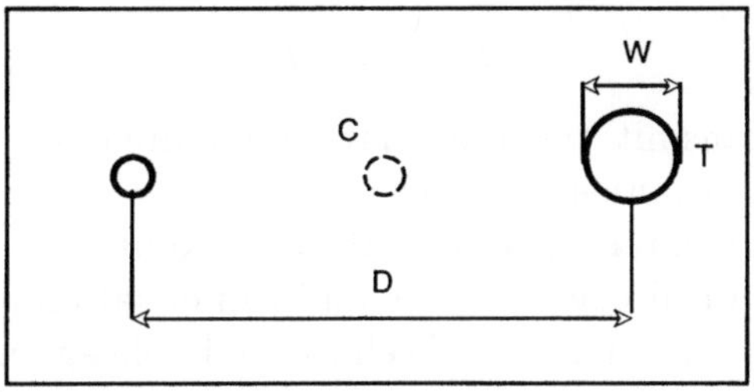

Figure 3.17. The targeting task performed by the HO. Designations: C is a cursor, T is a target, D is a distance of a working point movement, and W is a width of the target.

HO ability to realize quasi-time-optimal control during the performance of rather simple movements (Gidikov *et al.*, 1973; Kobrinskii *et al.*, 1971). Therefore, correctness of methods, based on some dynamic models of the HO, must be experimentally verified.

The second approach is based on the experimental study of robotic systems with their criterial estimate (as described above). Note that this effective method has a disadvantage that is associated with difficulties of an experimental study of system performances. To perform such a study, a telerobot (or its mathematical model) has to be already produced and, if a real telerobot is to be used in this research, it will be rather difficult to change something in its construction based on test results. The experimental design method application also leads to an increase in design cost and duration.

Thus, the development of a theoretical, computer-based approach is very desirable for progress in the telecontrol field. The third approach is based on behavioral studies of the HO. This approach has its origin in Fitts's results (1954), and a great number of studies now use the speed-accuracy trade-off known as Fitts's law. The principal result was experimentally obtained for human movements performed without any instrument.

Fitts's law basically describes a movement time T as a linear function of the *difficulty index* of aimed movement, defined as $I_d = \log_2(\frac{2D}{W})$, where W is the width of the target, and D is the movement amplitude (Fig. 3.17). The subject was then considered to be an information channel of limited capacity. This paradigm was discovered in subjects who were asked to move a hand-held stylus as fast as possible back and forward between two targets. Numerous studies have shown that this law holds for a wide variety of aimed movements produced under different environmental conditions, including single-joint wrist rotation, as well as multi-joint movements (Keele, 1968). Thus, it appears that Fitts's law reflects a fundamental property of human motor performance:

$$T = a + bI_d. \tag{3.9}$$

The time T to transmit the information required for task fulfillment is proportional to the channel's capacity.

In subsequent research, this estimate was modified to simplify a calculation with undetermined coefficients a and b of equation (3.9). An extension of Fitts's model was proposed by Welford and colleagues (1969), who considered two subtasks of a guidance process: travel, and positioning. The movement time can thus be modeled as

$$T = a_1 + b_1 \log_2 \frac{2D + W}{D_0} + b_2 log_2 \frac{D_0}{W}, \tag{3.10}$$

where in addition to designations of Figure 3.17, a_1 is a reaction time delay; b_1, b_2 are undetermined coefficients related to an information transmission rate for the subtasks; and D_0 is some point that separates the two parts of the task.

Following experiments showed that Fitts's result can also be extended to cases in which the HO performs operations using an instrument; that is, a manipulator (Kirkpatrick *et al.*, 1977; Verplank, 1975). This served as the beginning of many scientific investigations leading to the development of this third approach.

In the telerobotics field, the application of Fitts's estimate and its modifications to teleoperator design have it possible to evaluate and predict design performance. Kugath and Wilt (1974) noted that it was not practical to test experimentally the hardware for each new application, and that it would be useful to develop a means of analytical synthesis for a manipulator. They used Fitts's *difficulty index* to estimate a time-accuracy trade-off for the "peg-in-the-hole" test performed by different manipulators. More recently, this method was used and developed by Book and Hannema (1980), and a highly detailed description may also be found in a subsequent publication by Book (1985). To predict teleoperator properties, this methodology uses experimental data obtained for different values of robot parameters.

A multiple regression analysis has been applied to estimate the coefficients of the equation expressed in the form of the information transmission model (3.10) according to obtained experimental data. This was an attempt to quantify the relationship between the resultant indices of a MSS (the time completion targeting task) and each of three dynamic characteristics: natural frequency, Coulomb friction, and backlash. As a result, we were able to predict quantitative characteristics of a remotely controlled system for different combinations of its parameters. A disadvantage of this very developed approach is that the D_0 value in (3.10) is chosen by various means, which is reflected by the quality of the solution obtained.

In another approach, the initial dependence (3.9) was rephrased, based on the use of the natural logarithm (Cannon *et al.*, 1994):

$$T = \alpha + \beta \ln(\frac{D}{W/2}). \tag{3.11}$$

A special procedure was then proposed to determine unknown coefficients α and β, since response curves can be obtained from control theory models. We do not describe this approach in detail here, since it was already presented in Sheridan's book (1992). We note only that this procedure's implementation might be rather difficult, as well as insufficiently adequate, because of the necessity of involving dynamic models of complicated telerobotic systems.

3.4. Discussion

Two highly important problems of telerobotics were presented in this Chapter. The first was the problem of telerobotic system quantitative quality estimations. A system of unit manipulation operations and criterion estimates was introduced and methods of their use for simulation conditions and actual robot testing were described. It is important to note that almost all developed methods of IR testing may be applied to test remote manipulation systems. These system tests are, in fact, more comprehensive and include additional checks corresponding to more complicated operations that they perform. Also, a difference between various realizations of a control process will be much greater for telerobots than for IRs; consequently, the scatter of operation results will also be greater.

Methods for remote system study using simulation were presented based on typical task performance and the estimation of its quality with the help of a complex criterion. The latter includes indices of accuracy and task completion time. It was demonstrated that such an approach permits one to differentiate system quality, while changing different parameters of the telerobotic system. Among these, we can identify:
- different object gain (that is, a manipulator property);
- different skills of operators;
- different characteristics of the task to be performed (that is, task complexity).

Implementation of this approach with real telerobots leads to test stand organization, where unit manipulation operations are simulated and criterion estimate measurements are performed. This very productive approach is highly developed in all industrial countries and some results in this direction have been presented here.

The mnemonicability coefficient application to estimate remote manipulation systems quality was also described. This intermediate index gives some weighted estimation of system quality, and it is related, in a way, to criterion estimations of accuracy and task completion time. This index may be used at the preliminary design stage of a telerobot; that is, for a synthesis of remote control systems.

Lastly, the problem of remote manipulation system design is very closely related to the problem of system quality analysis. This tie is based on the use of estimates of accuracy and task completion time as criteria during system design. The most promising methods based on the application of Fitts's law were described above - an approach based on solving a targeting task of the type O_3 with the criterion estimation of the type C_4. While performing this typical task, the HO must move a cursor (or a robot endpoint) on a direct line from an initial point to a target of certain dimensions. By using this approach, methods have been developed that allow one to predict telerobotic system characteristics (Book and Hannema, 1980). Some methodical problems, however, are still unsolved, and therefore development and perfection of similar techniques in the future are desirable because the method has been shown to offer solutions for very important practical problems.

References

Book, W.J. (1985) Teleoperation arm design, Chapter 9, in R.C.Dorf and S.Y.Nof (eds.), *Handbook of Industrial Robotics. Intern. Encyclopedia of Robotics, Automation and Applications*. John Wiley & Sons, New York, pp. 138-157.

Book, W.J. and Hannema, D. (1980) Master-slave manipulator performance for various dynamic characteristics and positioning task parameters, *IEEE Trans. on System, Man and Cybernetics*, **Vol. 10 no. 11**, pp. 764-771.

Cannon, D.J., Thomas, G., Wang, C., and Kasavadas, T. (1994) A virtual reality based point-and-direct robotic system with instrumented glove, *Intern. J. of Industrial Engineering*, **Vol. 1 no. 2**, pp. 139-148.

Ceres, R., Martin, J.M., Calderon, L., Armada, M., and Pons, J.L. (1995) Photosensor tracks transients, *Industrial Robot*, **Vol. 22 no. 1**, pp. 21-24.

Elkind, J.J. (1964) A survey of the development of models for the human controller, in *Guidance and Control - 2*. New York - London, pp. 623-643.

Fitts, P.M. (1954) The information capacity of human motor system in controlling the amplitude of movement, *J. of Experimental Psychology*, **Vol. 47**, pp. 381-391.

Gamarnik, A.I., Ukhimov, V.V., and Shtromberg, R.M. (1989) A stand for manipulator testing, USSR author's certificate no. 1484692. *Bulletin of Inventions*, **no. 21** (in Russian).

Ganzen, V.A. and Granovskaya, R.M. (1965) On one quantitative estimation of configuration complexity, *Problems of Engineering Psychology*: Collection of scientific papers, **Vol. 2**. Leningrad, pp. 36-39 (in Russian).

Gidikov, A.A., Marinov, M.P., and Guevski, I.V. (1973) Bionic aspects of time-optimal control problem, in *Problems of Bionics*. Nauka, Moscow, pp. 402-413 (in Russian).

Hannaford, B., Wood, L., McAffee, D.A., and Zak, H. (1991) Performance evaluation of a six-axis generalized force-reflecting teleoperator, *IEEE Trans. on Systems, Man, and Cybernatics*, **Vol. 21 no. 3**, pp. 620-633.

Hill, J.W. (1979) Study of modeling and evaluation of remote manipulation tasks with force feedback, *Final Report, SRI Intern. Project* **7696**, March.

ISO (1990) *International Standard ISO 9283:1990(E)*. International Organization for Standartization, Geneva.

Keele, S.W. (1968) Movement control in skilled motor performance, *Psychological Bulletin*, **Vol.70 no.6**, pp. 387-403.

Kirkpatrick, M., Shields, J., N.L., Malone, T.B., Brye, R.G., and Frederick, P.N. (1977) Manipulator system performance measurements, *Mechanism and Machine Theory*, **2**, pp. 439-450.

Kobrinskii, A.A. and Kobrinskii, A.E. (1985) *Manipulation Systems of Robots*. Nauka, Moscow (in Russian).

Kobrinskii, A.E., Sergeev, V.I., Slutski, L.I., and Tyves, L.I. (1971) On the quality evaluation of manual control systems, *Mashinovedenie*, **no. 5**, pp. 3-10 (in Russian).

Kobrinskii, A.E., Sergeev, V.I., Slutski, L.I., and Tyves, L.I. (1974) Analysis of quality of manipulator manual control, *Cours. and Lect. CISM, Intern. Cent. Mech. Sci.*, **Vol. 1 no. 201**, pp. 203-212.

Kobrinskii, A.E., Statsenko, V.V., and Tyves, L.I. (1975) Mnemonicability level evaluation of the master-slave manipulator with disconcordance of the input and executing organs, *Mashinovedenie*, **no. 2**, pp. 3-7 (in Russian).

Kojinskaya, L.I. and Slutski, L.I. (1973) The role of the method of convolution in vector optimization, *Automation and Remote Control*, **Vol. 34 no. 3 part 2**, pp. 485-488.

Kugath, D.A. and Wilt, D.R. (1974) Problems in selection of design parameters affecting manipulator performance, *Cours. and Lect. CISM, Intern. Cent. Mech. Sci.*, **Vol. II no. 201**, pp. 169-189.

Liegeois, A. (1985) *Robot Technology, Vol. 7: Performance and Computer-Aided Design*. Prentice-Hall, Englewood Cliffs, NJ.

Medvedev, V.S., Leskov, A.G., and Yushchenko, A.S. (1978) *Control Systems of Manipulation Robots*. Nauka, Moscow (in Russian).

Ranky, P.G. and Ho, C.Y. (1985) *Robot Modelling: Control and Applications with Software*. IFS Publications Ltd., UK.

Sedler, M.Kh. and Martynov, V.S. (1987) A stand for robot testing, USSR author's certificate no. 1335448. *Bulletin of Inventions*, **no. 33** (in Russian).

Sheridan, T.B. (1992) *Telerobotics, Automation, and Human Supervisory Control*. The MIT Press, Cambridge, MA.

Sheridan, T.B. and Ferrell, W.R. (1974) *Man-Machine Systems: Information, Control, and Decision Models of Human Performance*. The MIT Press, Cambridge, MA.

Slutski, L.I. (1972) On quality criteria of manipulator manual control, *Mashinovedenie*, **no. 2**, pp. 28-33 (in Russian).

Slutski, L.I. (1977) A stand for manipulation system quality evaluation, USSR author's certificate no. 580113. *Bulletin of Inventions*, **no. 42** (in Russian).

Slutski, L.I. (1980) Methods of robotic system qualimetry, *Proc. of Kazakh Seminar on the Machine and Mechanism Theory*, **Vol. 3**: Collection of scientific papers. Kazakh State University, Alma-Ata, pp. 65-73 (in Russian).

Slutski, L.I., Vorobiov, P.I., Shafigov, U.M., Vasil'anov, V.P., Fel', A.K., and Ivanchenko, V.S. (1989) An optoelectronic measuring device for accuracy characteristic determination of industrial robots, USSR author's certificate no. 1519073 (in Russian).

Statsenko, V.V. and Tyves, L.I. (1985) Towards creation of a rational control system of a manipulator, *Investigation and Solution of Applied Mechanics Problems on Computers*: Collection of scientific papers. Nauka, Moscow, pp. 65-73 (in Russian).

Verplank, W.L. (1975) Research on remote manipulation at NASA/AMES Research Center, in T.Sheridan (ed.), *Performance Evaluation of Programmable Robots and Manipulators. Reports of a Workshop*. U.S. Department of Commerce, NBS SP-459, Annapolis, MD, pp. 91-95.

Vukobratovic, M. (1976) *Walking Robots and Antropomophic Mechanisms*. Mir, Moscow (in Russian).

Warnecke, H.J. and Schiele, G. (1985) Performance characteristics and performance testing of industrial robots - state of the art, in: *Robot Technology and Applications: Proc.*

of 1st Robotics Europe Conf., Brussels, 1984. Springer-Verlag, Berlin, pp. 5-17.

Welford, A.T., Norris, A.H., and Schock, N.W. (1969) Speed and accuracy of movement and their changes with age, *Acta Psychologica*, **Vol.30**, pp. 3-15.

Wilt, D.R., Pieper, D.L., Frank, A.S., and Glenn, G.G. (1977) An evaluation of control modes in high gain manipulator systems, *Mechanism and Machine Theory*, **Vol.12**, pp. 373-386.

Wodzinski, M. (1987) Putting robots to the test, *Robotics Today* , **June**, pp. 17-20.

CHAPTER 4

CONTROL METHODS IN THE SYSTEMS "OPERATOR-REMOTE MANIPULATOR"

4.1. Kinds of Remote Control Methods

A general classification of remote control methods is presented in Figure 4.1. Note that it does not include manual control methods when direct force interaction occurs between the HO and a robot, which includes different exoskeleton-type devices as well as robotic teach-in systems. All these cases relate, in fact, to manual robotic control, and they are not, at the same time, remote control methods. 117

In our classification, it is worth mentioning, first, myoelectric control from muscles, which was primarily used to move prosthetic devices and which A.Kobrinskii and his colleagues (1960) pioneered. In addition to its application to the control prosthetic devices, this method can also be used to control remote manipulators. A number of studies have been carried out in this direction following the works of R.Tomovic (1960; 1969). This method offers the HO some options in controlling equipment, including:

- increased control speed, as compared with ordinary manual controls, due to a shorter time reaction in the myoelectric control channel on 40...80 msec;
- the use of the myoelectric channel allows the HO to control a greater number of robot DOFs and auxiliary devices;
- myoelectric control principles, having been learned by the HO, are not disturbed by, and can function in, any very complicated conditions (even if the HO has no opportunity to move);
- assurance of non-fatiguing control because the method does not require a great physical load on the HO.

In recent years, new attempts have been made to advance this method based on current technical facilities (Farry and Walker, 1993). However, the low accuracy of the method does not permit it to be widely applied to control telerobots. HO arm motions are routinely used for this purpose. During these motions, the HO uses his/her arm motions to operate different kinds of input devices. The kind of input device depends on the specific control method in use.

To date there are four such methods used for a number of applications. The first one, the on-off control method, which is a widely used technique,

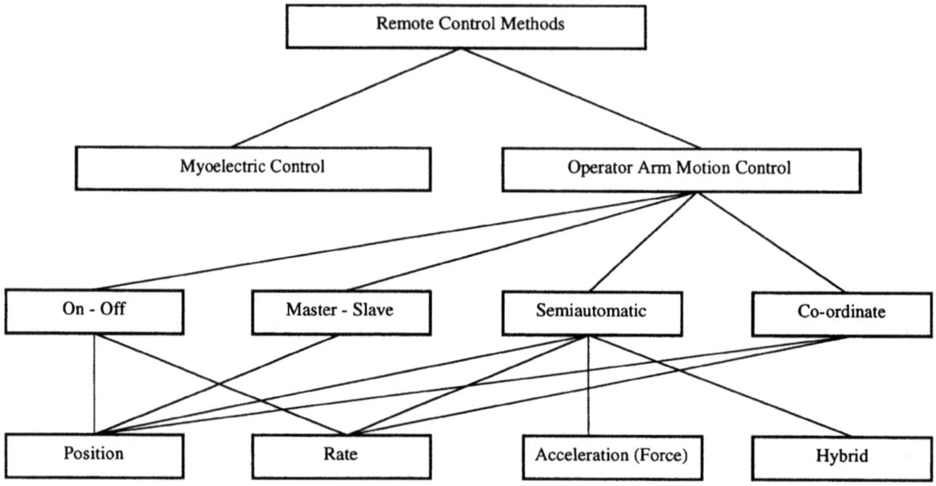

Figure 4.1. Classification of remote manipulation systems.

is especially used in robotics: a set of on-off pushbuttons is employed, each corresponding to different DOFs of a robot.

Another control method involves master-slave systems (MSSs). A principal feature of this method is that the executing organ (EO) geometry is similar at a certain scale to an input device (ID) geometry.

The most useful current approach is a method of semiautomatic control (SAC). A principal characteristic of this kind of remote control is the use of an ID with kinematics that is dissimilar to that of the EO geometry.

The co-ordinate control is also used for a number cases. In this case, a control is realized, as in the case of the on-off control, by means of pushbuttons; however, this control is used, not for direct drive control, but to organize a robot EE shift according to different coordinate axes.

It is important that, as a rule, all these control methods be used in telerobots in combination. Some DOFs of a robot can then be controlled by one method, and other by the second or third method. These methods can be united with sensing systems (which are briefly presented in this book). A necessity to combine these methods arises frequently to ensure required quality characteristics.

The above methods are used to control telerobots according to requirements of their EE position, velocity, acceleration, and force. The control methods are described here from the viewpoint of their quality, which is considered according to the criteria introduced in Chapter 3. First, however, we would like to present material that is generally related to existing types of IDs and their features.

4.2. Input Devices (IDs) of Semiautomatic Control (SAC) Systems

In the following, only IDs of SAC systems are described, since they are IDs of the most general type and might be used to control any kind of telecontrol systems, including MSSs.

Among these kinds of IDs (for example, keyboards), hand controllers (HCs) are used to introduce smooth input signals to flexible control drives of different telerobot DOFs. A feature of HCs of the SAC systems is that the same device is used to set different control modes of a robot's work: position, velocity (rate), and acceleration (force). The vector produced by the HC is then proportional to a shift vector of a manipulator gripper, to a velocity vector of the gripper, and determines forces and moments applied to an object, respectively.

Another known kind of robotic HC application is when each separate movement in a certain HC joint is used to control a specific DOF of a manipulator. Examples of this implementation are presented in Vertut and Coiffet (1984), where a *syntaxer*, for instance, was described as such an ID. The kind of control obtained in this case is one that is a certain intermediate between an on-off control and a proper semiautomatic one, and it is mentally difficult for the HO, because he/she solves inverse kinematics in his/her head, which is not entirely effective.

4.2.1. HAND CONTROLLERS

Hand controllers are sometimes called " masters" in literature, regardless of the type of system with which they are used. A classification of these principal types (Fig. 4.2) must include as decisive factors a number of controlled axis and kinematics of the mechanism, which is basically a HC. Types of kinematic pairs and sensors may also be considered as classification signs; however, these signs are secondary. The means for a force reflection may also be included in these devices, which would allow the HO to improve his/her manipulations with remote physical objects or interactions with objects of "virtual" environment.

A number of different requirements are demanded of HCs for robotic remote systems. Among them, the following criteria should be noted (Fischer *et al.*, 1990):
- volume of operation (there is usually difficult in optimizing this index when characteristics of a control system are constant, because the principal demand of a small value for this volume in order to minimize HO fatigue does not ensure precise teleoperations);
- inertial effects (the demand to minimize this criterion is based on our own results (see Chapter 5), and those of other authors, related to maximi-

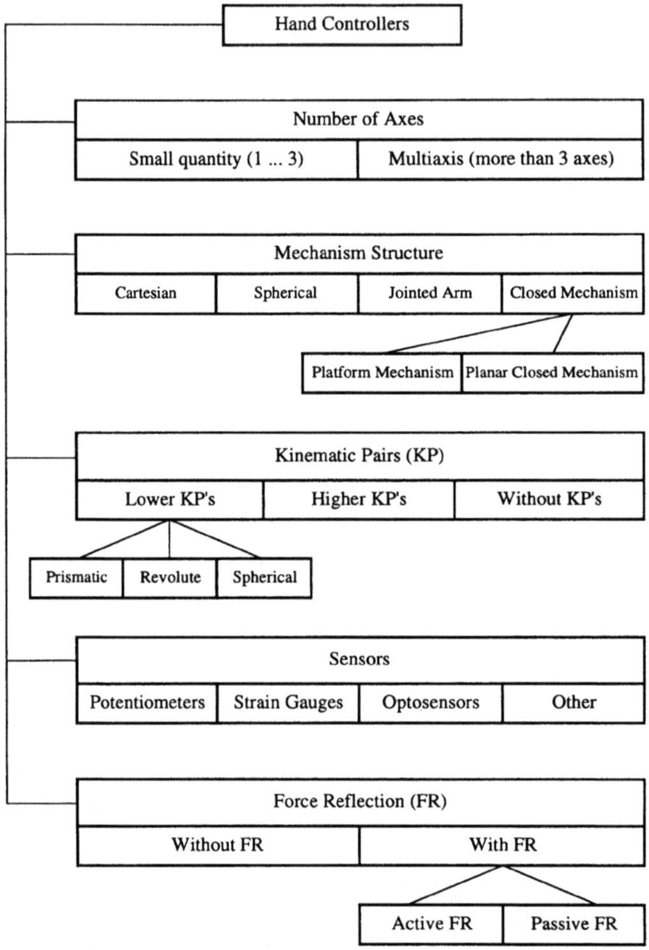

Figure 4.2. Classification of hand controllers (HCs).

zing transparency of telerobot control systems (see, e.g,, (Handlykken and Turner, 1980));
- cross coupling effects (when control signals for certain controlled coordinates are produced, such signals for other controlled coordinates do not need to be produced; this demand of an absence of cross-coupling control coordinates sometimes leads to great difficulties while designing these HCs - especially multi-axes devices); etc.

These criteria depend on a number of structural characteristics of HCs. Concerning types of basic HC kinematics, we would like to note that HC structures almost repeat existing structures of IR mechanisms. For example, spherical, Cartesian, and jointed arm input control devices correspond to the classification of robot structures in the book (Klafter *et al.*, 1989). We would add to this classification of conventional opened robotic mechanisms a group of closed kinematic chains as well. Among these we can distinguish two principal types: proper platform mechanisms; and, platform mechanisms based on closed planar kinematic chains.

As previously noted, all mentioned mechanism types may be combined while creating multi-axis HCs.

Among the exceptions to this classification, we would distinguish special type of IDs used in the computer techniques that may be also utilized to control robotic devices. This is the ordinary computer mouse, which may be applied to select the target location when a robot graphic simulation is in use. As an example of such an experience, we point out the paper (Nechyba and Xu, 1995), where the HO used a mouse to mark a target on a graphic display that provided two views of the space station truss structure. Less usual, but preferable for robotic applications, is the 3D-mouse, which is described in (Burdea and Coiffet, 1994).

HCs of **Spherical** *Structure*

As in robot mechanics, the term "spherical structure" does not always mean the direct use of only spherical joints. There are cases when the entire basic joint is in use, but sometimes only separate elements of its structure are used to set control signals.

First, all possible types of joysticks are related to this type of HCs. The main kinematic structure of these joysticks is usually used to set two-coordinate signals to control motion in a plane (Fig. 4.3). This is applicable to control robot motions in horizontal planes (e.g., for Cartesian robots) and for different other kinds of equipment (e.g., cranes). In this case, other robot DOFs are controlled with the help of different pushbuttons, which are usually placed on the joystick hand-grip. A typical example of these devices is a handgrip shown in Figure 4.4.

Ordinary potentiometers are usually used as position sensors for these joysticks. However, simpler touch sensors (e.g., limit switches) are also in use instead of potentiometers. While controlling the HC, the HO does not produce a smooth control law but sets a relay (on-off) control when he/she moves a handle in extreme positions for each controlled coordinate.

Gimbaled HCs are also a particular case of spherical IDs. In this case, the spherical joint is implemented with the help of three-revolute joints with axes that are intersected at one point. The gimbaled mechanism may be used to set linear displacements of the control handle (Saenger and Pegden, 1973), and may be used to set rotational motions for robot gripper orientation (Siva *et al.*, 1991) (see below - Fig. 4.7).

Figure 4.3. Joystick with over-drive gears together with potentiometers. Courtesy of Sakae Tsushin Kogyo Co.

A spherical, robot-like structure of HC was also used in an advanced telerobotic system developed in the Jet Propulsion Laboratory (Bejczy, 1994). To set spatial coordinates of a robot hand, a two-DOF basic mechanism is added (Fig. 4.5) by a translation axis. A resultant three-axis mechanism is connected to a gimbaled hand-grip that sets robot orientation motions.

A similar solution was used in our project described below in Chapter 6.

It can be seen that this kind of HC is very preferable for use in telerobotics. A major reason is their relatively small working volume, which is important in many applications.

Figure 4.4. A control grip which incorporates two-axis control systems, two pushbutton switches, and the trigger switch. Courtesy of Measurement Systems Inc.

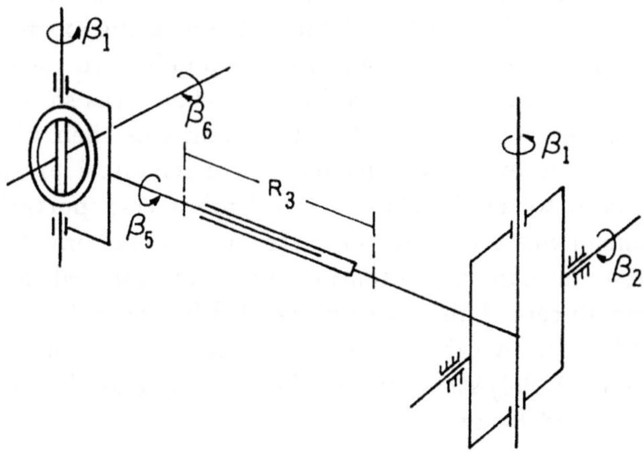

Figure 4.5. Kinematics of the multi-axis HC of the Jet Propulsion Laboratory, Pasadena, CA, USA. From (Bejczy, 1994); ©American Institute of Aeronautics and Astronautics.

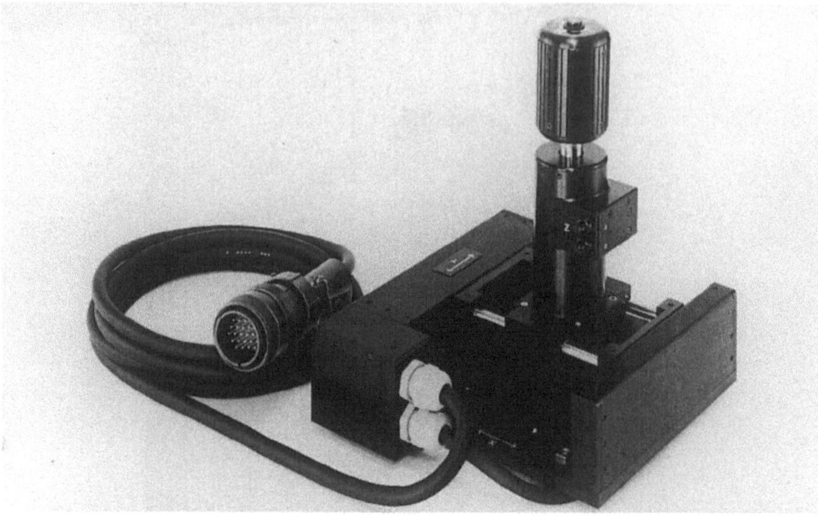

Figure 4.6. Six-axis HC based on Cartesian structure. Courtesy of Sakae Tsushin Kogyo Co.

Cartesian *HCs*

A proper Cartesian HC is shown in Figure 4.6. The Cybernet System Co. experience in a HC design (Jacobus *et al.*, 1992) testifies that such a scheme ensures maximum HO gripping force and control. An impressive specification of requirements on robotic HC design is also presented by Jacobus *et al.* .

Control vector setting in Cartesian coordinates may be carried out in different ways. Siva and colleagues (1991) give us some examples, and propose implementation of control signals corresponding to Cartesian coordinates with the help of either a serial linkage scheme (as shown in Fig. 4.6) or a parallel linkage scheme. The latter version was their ultimate choice (Fig. 4.7) because it exhibits homogeneous mechanical properties. This structure was used to set signals to control robot transportation DOFs. A gimbal mechanism was attached to the structure to set orientation motions of a robot wrist mechanism. Potentiometers were located in the HC axes as position transducers. This device was called by the authors the CARTESIAN mini MASTER ARM (CmMA). It may serve as an example of the use of platform (parallel) structures in the field. A general approach to their use will be described below.

An advantage of the proper Cartesian HCs is that there is no need to additionally transform their coordinates before using them in the control system; that is, the system software is simplified. Control in Cartesian coordinates is also more understood by the HO, and this increases telerobot

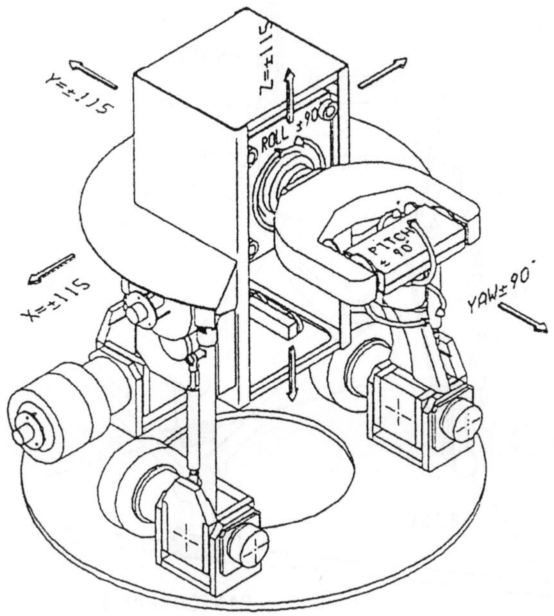

Figure 4.7. HC based on a parallel linkage scheme with an attached gimbaled mechanism to set orientation motions. From (Siva *et al.*, 1991).

productivity.

Jointed Arms

This kind of HC is very used in robotics. For instance, all master arms of MSSs relate to this type of control device. Therefore, they are described many times and in detail in the literature (e.g., (Vertut and Coiffet, 1985)). One recent application is a 6-axis force reflecting HC used by NASA to built the Flight Telerobotic Servicer (Andary and Spidaliere, 1993). Thanks to modern computer technique achievements, these devices may be used not only in MSSs, but also for SAC systems, which demand corresponding coordinate transformations (Jansen *et al.*, 1992).

Platform Mechanisms

The application of platform mechanisms in robotic executing systems may be found to K. Hunt's papers (e.g., (1983)), which presented the most detailed description. Proper platform mechanisms have versions of both executing and input robotic devices. There are two possible approaches to building these mechanisms. The first is illustrated by the paper (Hollerbach and Lokhorst, 1995). Such a HC presents a spatial platform mech-

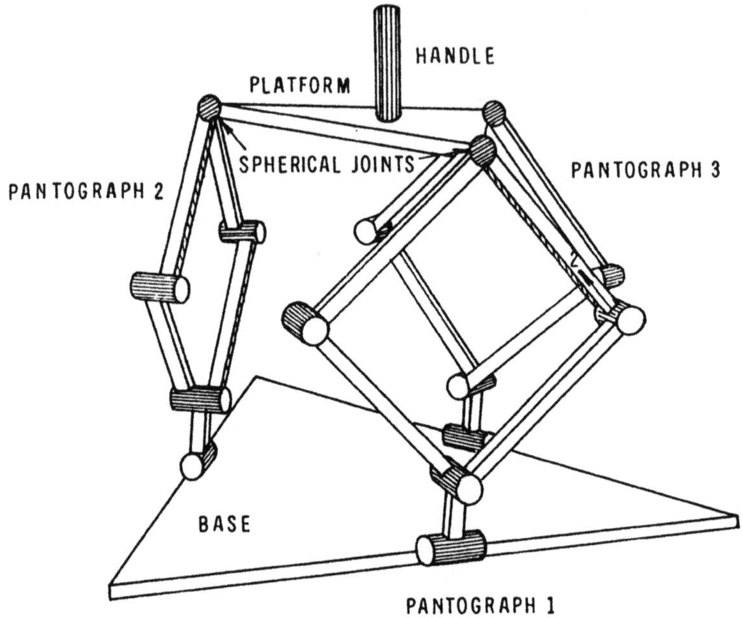

Figure 4.8. Layout of the platform type HC.

anism (Fig. 4.8) with a hand-grip placed on the moving platform. The platform is attached to a base, for example, by three pantograph mechanisms (Collins and Long, 1995). Potentiometers are placed in pantograph joints to measure mechanism movements. The use of these mechanisms as telerobot HCs is based on the conclusion that their mechanical stiffness makes them more attractive for force reflection than serial-chain structures. One practical example of such a device is presented in Figure 4.9. It was used to implement the hybrid position-force control for teleoperations with time delays (Reboulet *et al.*, 1995). Such an input device was found very sensitive and capable of rendering better motion sensations.

The second approach to building these devices is associated with spatial mechanism creation by means of two kinematic chains that connect a platform element to the frame. The platform element can be made in pen form (Iwata, 1993), which does the HO work more conveniently (see also, (Burdea and Coiffet, 1994)).

Concerning the application of closed planar mechanisms in robotic input devices, we refer, first, to our results described below (Fig. 4.25). This system ID was performed just like a HC recently developed by Long and Hao (1995). Of particular note in their paper is that proper in-parallel mechanisms have more singular configurations than their serial-chain coun-

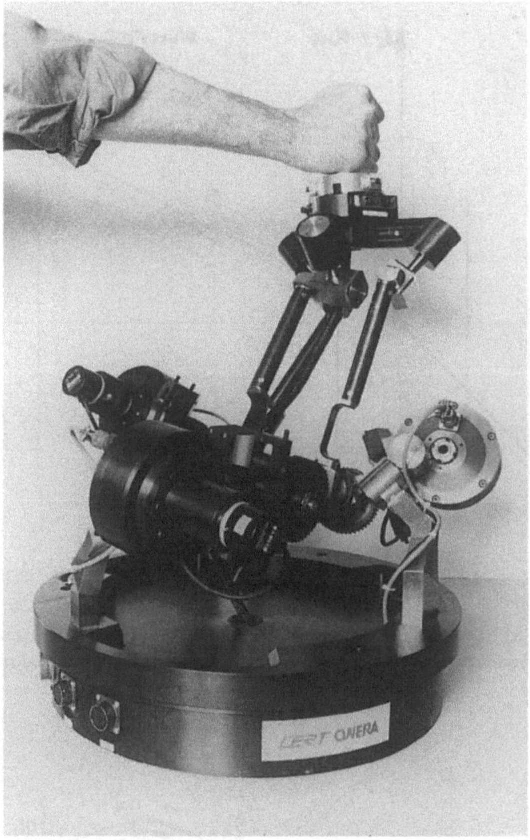

Figure 4.9. Platform type HC of CERT (Centre d'Etudes et de Recherches de Toulouse), France. Courtesy of C. Reboulet.

terparts, which further complicated all considerations of force-reflection.

According to terms of mechanism and machine theory, kinematic pairs, which are used in HCs as well, are distinguished as "lower" and "higher." Lower pairs are characterized by contact between surfaces of kinematic pair elements. Revolute, prismatic, cylindrical, spherical and other kinematic pairs are related to this kind of pairs. Pairs where a contact of elements occurs on lines or at points are called the higher kinematic pairs.

Lower kinematic pairs, especially those that revolute or prismatic, are usually used in HC constructions. When using only revolute pairs, a spatial vector shaped by the HC is set (as shown in Figure 4.5) in spherical coordinates (Fig. 4.10), or as a sum $\vec{X} = \vec{r}_1 + \vec{r}_2 + \vec{r}_3$ of certain components, in the case of a HC in a form of a jointed arm (Fig. 4.11). As position sensors, rotating potentiometers are usually used in these devices.

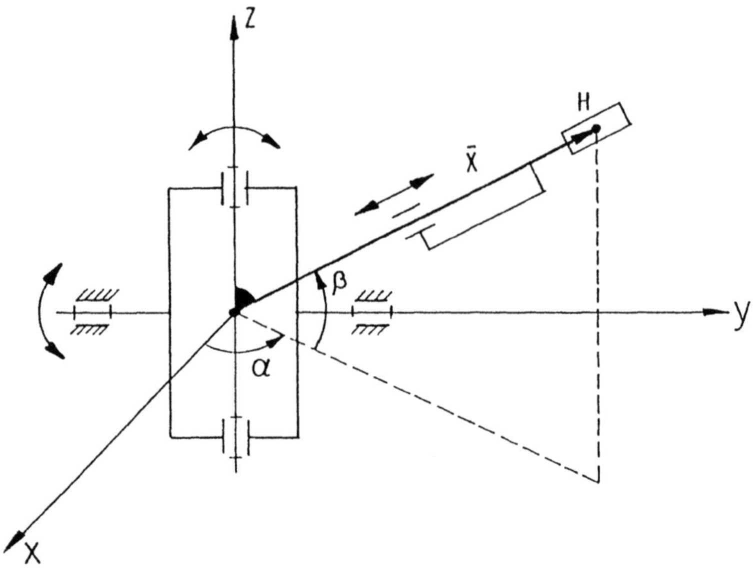

Figure 4.10. An input vector is set on the HC whose structure corresponds to spherical coordinates using two revolute and one rectangular pairs (H is a control handle).

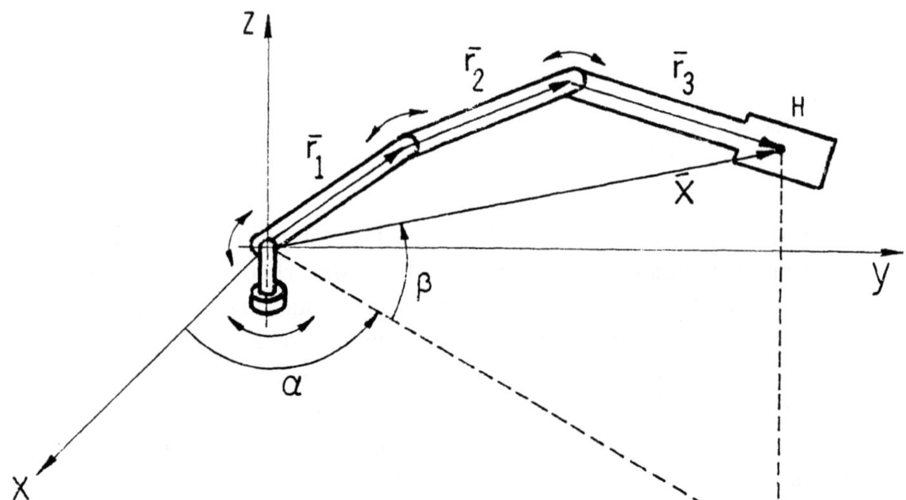

Figure 4.11. An input vector is set on the jointed arm HC with four revolute kinematic pairs (H is a control handle).

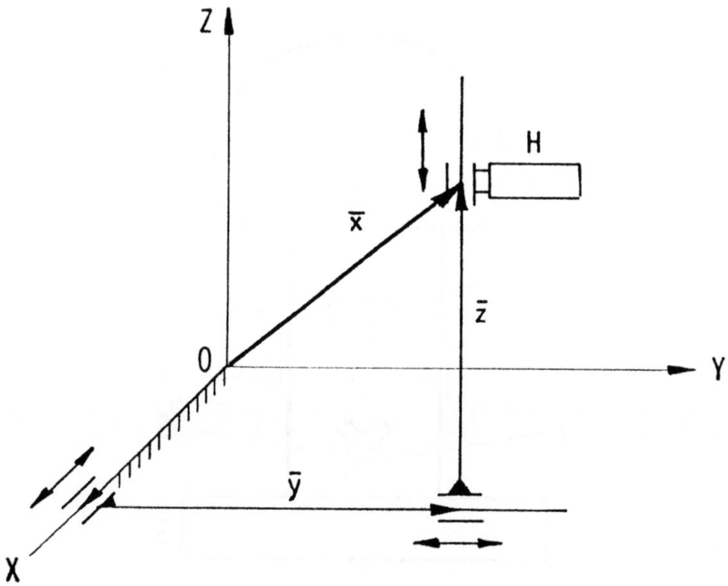

Figure 4.12. An input vector is set on the HC whose structure corresponds to Cartesian coordinates using three rectangular pairs (H is a control handle).

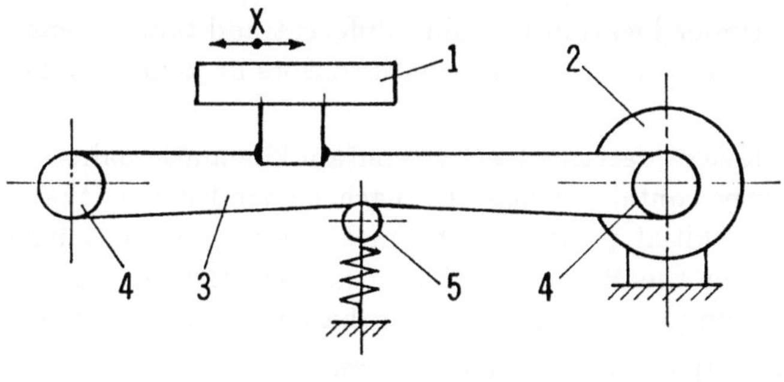

Figure 4.13. Transmission from a linearly moved link of the HC to a sensor. Designations: 1 is a movable link, 2 is a sensor (e.g., a rotating potentiometer), 3 is a thin flexible cable, 4 is a pulley, and 5 is a device for creating tension on the cable.

When using prismatic pairs, a control vector $\vec{X} = \vec{x} + \vec{y} + \vec{z}$ is set in a Cartesian frame (Fig. 4.12). In this case, linear potentiometers or rotating potentiometer with a corresponding transmission (see, e.g., Figure 4.13) are suitable as sensors to measure components of the vector to be set.

These opened kinematic chains are used as a basis for the construction of the HC. The orientation vector of the robot EE may be added to the

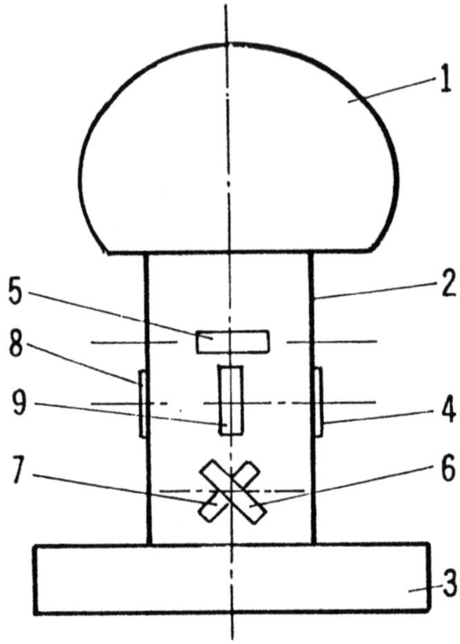

Figure 4.14. Isometric joystick. Designations: 1 is a grip; 2 is a bar; 3 is a base; and 5-9 are strain gauges. From (Wurll and Fritz, 1985).

vectors of gripper location by using different additional means (buttons on a handgrip (Fig. 4.4) or linkages with sensors in joints, as shown in Figure 4.5).

All the devices described above contain kinematic pairs of the 5th class, and frequently contain springs for each controlled axis that will return a handle to its initial position. The means for heavy HC balance are also used to lighten the HO work. These means may be either a mechanical counterbalancing or counterbalancing based on the use of servodrives.

These examples do not exhaust possible HC implementations. There is, for instance, an interesting class of HCs that does not contain absolutely kinematic pairs in its structure. A large group of these HCs are based on measuring small deformations of HC flexible elements. Since we are dealing with small deformations and their measurement, it is clear that strain gauges are usually used here as sensors. As in the above-described cases, it is possible to set both 3D- and 6D-vectors with the help of this approach. An example of a HC that uses this principle may be seen in Figure 4.14. This device has a joystick with a grip (1), but its bar (2) is rigidly connected with a base (3). The bar is a tube, and strain gauges (5-9) are placed on it that measure forces applied by the HO. For example, an axis moment is measured based on bar deformations with the help of two pairs of strain gauges (6, 7), which are glued at an angle of 45° to the bar axis.

Such devices are frequently called **isometric**. They have a relatively small working volume, but their common disadvantage is they are cross-coupled between different axes of a HC. Cross-coupling can cause reduced controllability and inaccurate positioning and force reflection. Much more perfect devices belonging to the class of isometric HCs were developed in USA (SRI, Draper Lab etc.) as early as the end of seventies. In order to explain the functioning, let us take a modern instance made by the Institute of Robotics and System Dynamics, Germany.

In this device (Heindl and Hirzinger, 1986), a basic idea of a 6-DOF control ball was developed and arranged so that the measurement center of a 6-component force-torque sensor approximately coincides with the center of the (plastic) hollow ball, into which is integrated (Fig. 4.15).

In this construction, a reference numeral (2) indicates a force/torque sensor that is integrally connected to the inner wall of a hollow sphere (10). The latter is made of a light material such as a rigid and hard plastic material. The circumferential ring (25) of sensor (2) is connected via four spokes (24) to a load pick-up plate (23). The lower surface of the plate (23) is connected via four supports (22) extending at right angles to the spokes (24) and being integral with a base piece (21). Strain measuring gauges are applied to the upper and lower sides of the spokes (24) and to the vertical sides of the posts (22) so as to produce voltages proportional to deformation or bending of the pick-up plate relative to a fixed inertial platform.

Except for common drawbacks usual for isometric control devices, this HC is characterized by the high cost and complicated hand manufacturing of mounting 16 strain gauges, as well the high cost of the necessary amplifier electronics connected to the strain gauges.

The ball idea was therefore redeveloped in 1986 into the SPACE MOUSE cap (Fig. 4.16) based on the principle of optoelectronic measurement. The basic construction (Dietrich and Plank, 1988) contains a light-emitting device in the form of a light-emitting diode (2-1), provided on a mounting device (2) (Fig. 4.12). A slit diaphragm located within a cylindrical ring (3) is allocated to diode (2-1) at a fixed, predetermined distance. Linear position-sensitive sensors (4-1) are used here as receivers of optical signals, as in our devices for robot testing (Chapter 3). This sensor, attached to a cylindrical ring (4), is in a vertical position relative to the direction of the slit of the respective slit diaphragm. In the optoelectronic array of this device, the six light-emitting diodes (2-1 through 2-6) are positioned on a side surface of the part (2). Adjacent slit diaphragms in ring (3) are shown in a staggered array at angles of 90° to each other. A minimum of six

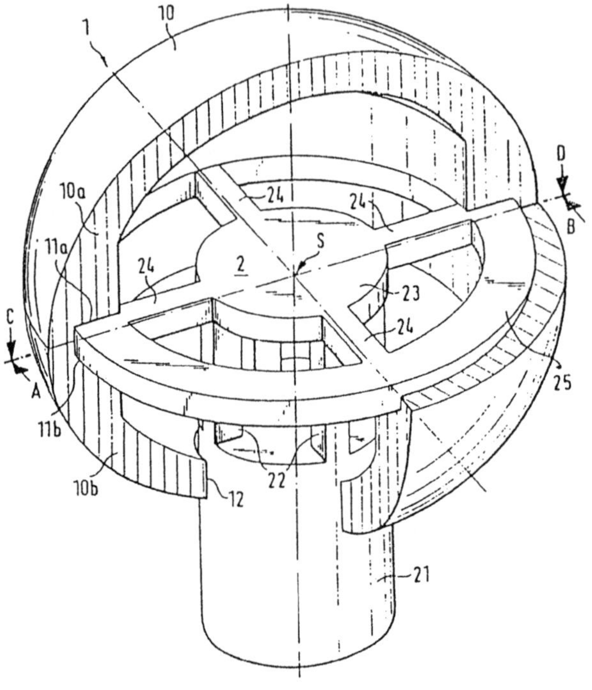

Figure 4.15. A perspective view of a hand grip-sensor for programming movements of a robot. The following principal designations are used: 2 is a force/torque sensor; 10 is a hollow sphere; 22 are posts; 24 are spokes, and 25 is a circumferential ring. From (Heindl and Hirzinger, 1986); reproduced here by kind permission of the US Patent and Trademark Office.

position-sensitive sensors (4-1 through 4-6) are fixed to the inside of plastic sphere (1). The sensors are movable together with sphere (1) relative to the respective light-emitting devices and slit diaphragms. By moving the sphere (1), the HO sets six position voltages of sensors (4-1...4-6) that are digitized and fed to a microcomputer to simply calculate corresponding six displacements and angular rotations.

Except for the SPACE MOUSE, this idea was implemented in the optoelectronic ball (Fig. 4.17), which has flown in space as well been used on the ground for teleoperating the real space robot ROTEX and for remotely programming it in the virtual environment on Earth in the frame of the "tele-sensor-programming approach" (Hirzinger, 1993).

This 6D space mouse idea was used in a new design of control panel KCP (Fig. 4.19) of the IR "KUKA" control system KRC1, which has been seen by the author at the Biennial International Machine Tool Fair, October 1996, in Milan and mentioned in (Kochan, 1996). This ergonomically designed

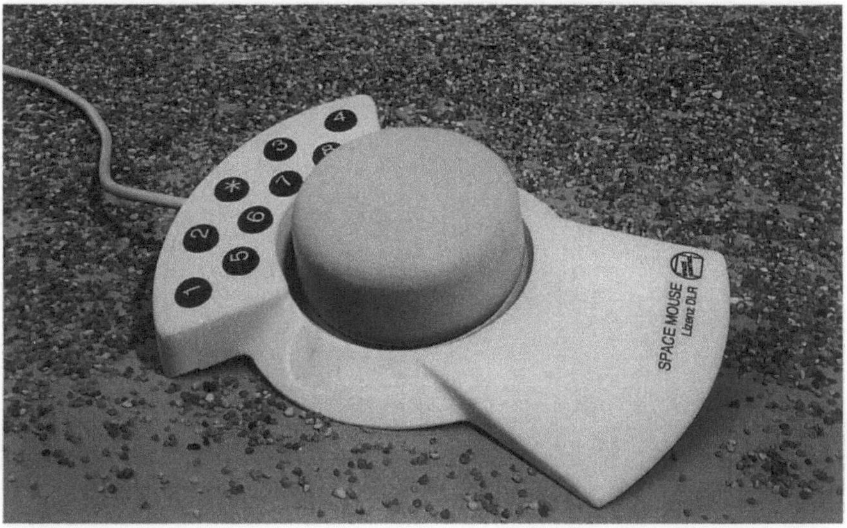

Figure 4.16. The SPACE MOUSE cap that presents an optoelectronic ball for multi-DOF control. Courtesy of G. Hirzinger, Institute of Robotics and System Dynamics, Germany.

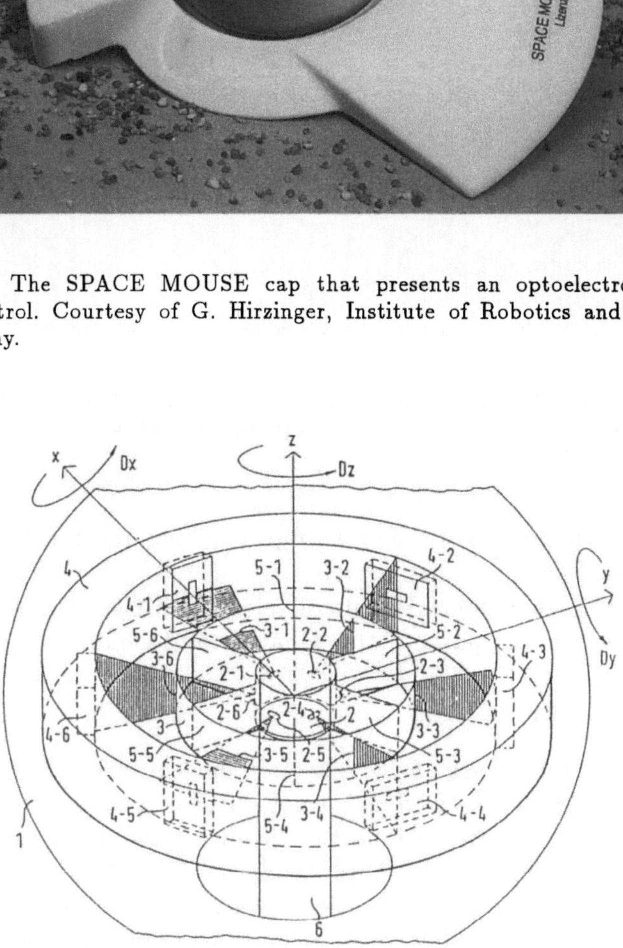

Figure 4.17. A schematic perspective view of an optoelectronic array housed in a plastic sphere. The following principal designations are used: 2-1,...,2-6 are light-emitting diodes; 3 and 4 are cylindrical rings; 4-1,...,4-6 are linear position-sensitive sensors. From (Dietrich and Plank, 1988); reproduced here by kind permission of the US Patent and Trademark Office.

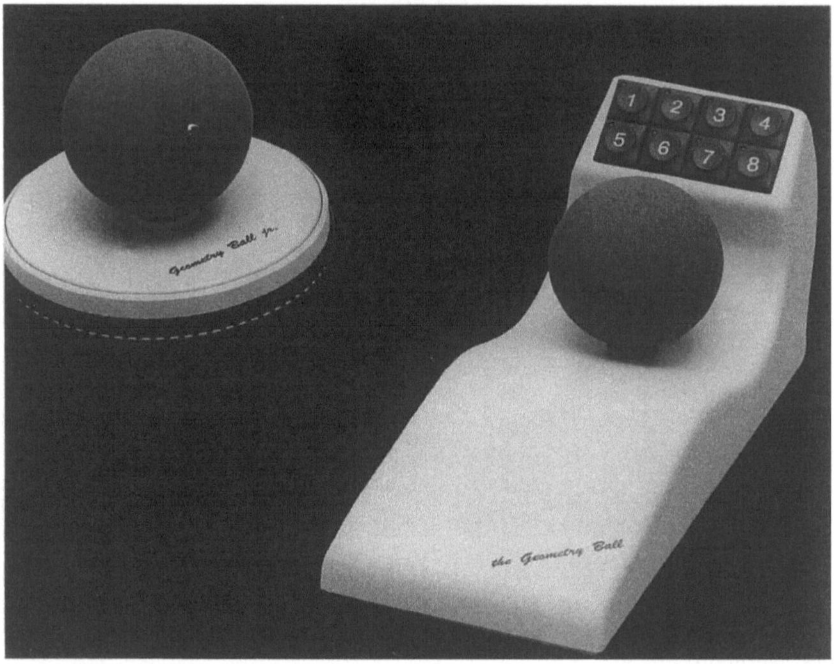

Figure 4.18. The Geometry Ball enables the control of 3 rotational and 3 translational DOFs with one hand. Courtesy of G. Hirzinger, Institute of Robotics and System Dynamics, Germany.

hand programming unit is used for robot teaching and constitutes the man-machine interface.

Note that several special kinds of telerobot HCs are also described below, in corresponding parts of this Chapter and Chapter 6.

4.2.2. FORCE REFLECTION

Let us now briefly describe force reflection devices, which are usually placed on or near a HC. Force feedback to the HO is a very important issue of remote control. First, force reflection from a slave manipulator causes an essential increase of remote manipulation quality (Hannaford *et al.*, 1991). Second, there is a circle of operations based on force interactions between a robot gripper and its environment (for instance, operations of a swivel-mounted cover opening, peg inserting into a hole, etc.) that cannot be performed without using sensor information about these interactions. Present-day HCs are therefore often provided by special devices for this purpose (Bejczy, 1994). There are two principal approaches to implementing this technique. The first, which is the traditional one, consists of placing corres-

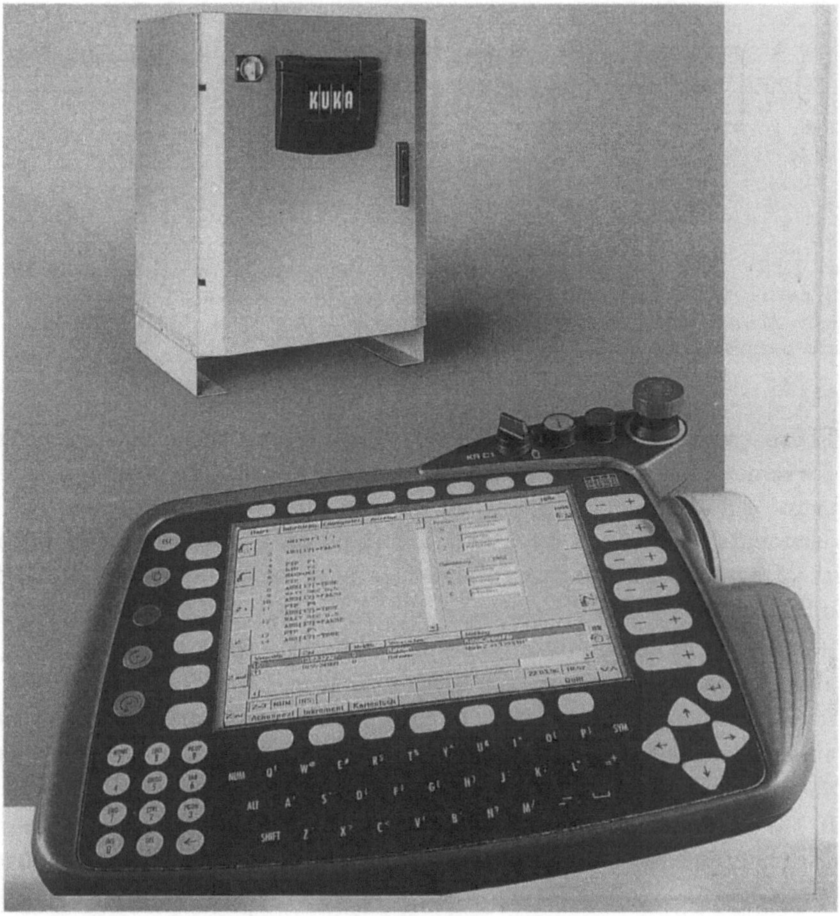

Figure 4.19. Control panel KCP of the IR "KUKA" with the 6D space mouse (located on the right side of the panel). The KCP features an 8-inch full graphics color display, a membrane keyboard, and the operator control elements as well. Courtesy of KUKA Roboter GmbH.

ponding motors in HC joints that produce forces proportional to sensing signals. This approach is sometimes called "active force reflection." Another approach is to place clutches in HC joints that brake the joints according to signals from a slave. This kind of force reflection is called "passive reflection." Generally, this kind of feedback is called kinesthetic force reflection, and is perceived by the HO through his/her proprioreceptors. For the first kind of sensing, brushless DC motors are usually used that provide good quality teleoperations. An example of such a control system is shown in Figure 4.20.

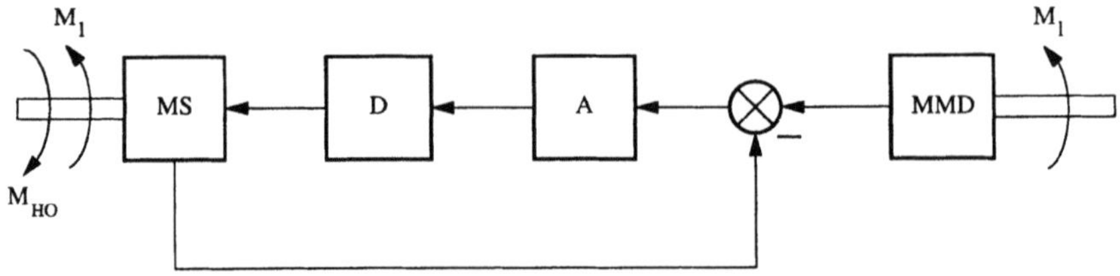

Figure 4.20. Scheme of active force reflection. Designations: MMD is a moment measuring device, D is a drive motor, A is an amplifier, MS is a moment sensor, M_l is a load moment, M_{HO} is a HO moment. Adopted from (Kuleshov *et al.*, 1986); reproduced here by kind permission of Mashinostroenie, Moscow.

There are many different problems in designing these sensing systems, but for considerations of this volume's length, we will only mention here the problem of matching power and information bandwidths between the system elements (Fischer *et al.*, 1990) and mention the most general problem - that of system transparency - which is now becoming a very important issue (Lawrence, 1992).

However, not all kinds of telerobots can use this kind of sensing; for example, it is impossible when using the isometric HC as a system ID. Moreover, while using the traditional approach, there are some insensitivities in the channel of robot force reflection that prevent the HO to feel small forces (less then 500 g); that is, while performing fine assembly operations.

In this case, another approach may be applied that consists of information transference to skin receptors by means of so-called skin displays. It has been learned, for example, that local vibrotactile irritations have the most informative characteristics, the HO does not have the ability to adapt to them during sustained activity, and his/her fatigue does not increase rapidly. Vibrotactile irritation parameters that can be used and changed to transmit information to the HO include: amplitude, frequency, irritation persistence, vibrator locations, etc. Thus, robot force sensor signals are coded and sent to a matrix of vibrators placed on an input handle. In the example shown in Figure 4.21, this matrix is fixed on the human hand that controls a master handle.

By experimentally studying this device (Efremov, 1978), it was established that it permits the HO to more exactly regulate (dose) grasping force. Since this takes place in a range of small forces, dosing errors do remain below 10%.

This sensing system testing was realized in Prof. Kobrinskii's Laboratory in the Institute for the Study of Machines, and the use of the complex criterion J (described in Chapter 3) was applied. Tracing different planar

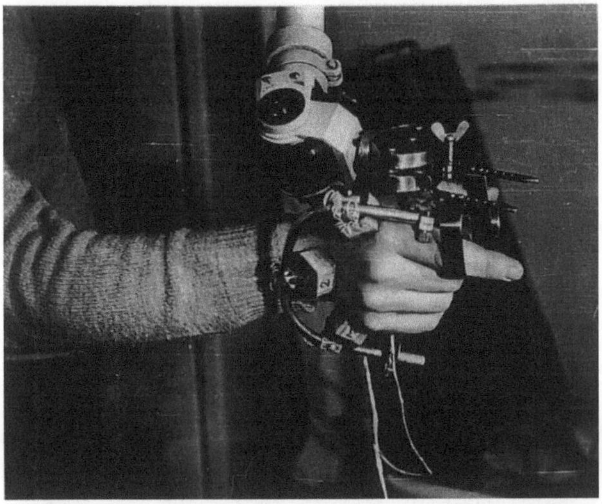

Figure 4.21. Master manipulator with vibrator matrix. Courtesy of A. Schneider.

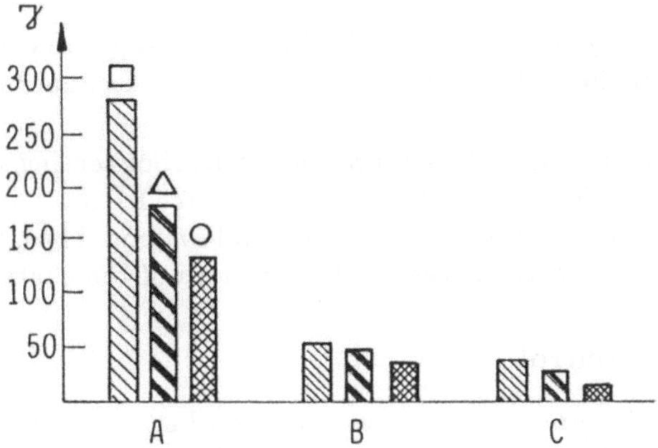

Figure 4.22. Results of experiments on reference trajectory remote tracing by means of only force feedback (A), only visual feedback (B), and both visual and force feedbacks (C). J is the complex criterion (3.3), types of trajectories are designated on the graph. Courtesy of A. Schneider.

prominent reference curves was one of the testing operations. A slave arm was provided by a two-component measuring device (Fig. 4.21), signals of which were transformed into vibrotactile irritations. The results of this investigation indicated that, by using the described sensing system, the HO was able to perform a set task even without visual feedback; the results of this performance, however, were not satisfactory (Fig. 4.22). The use

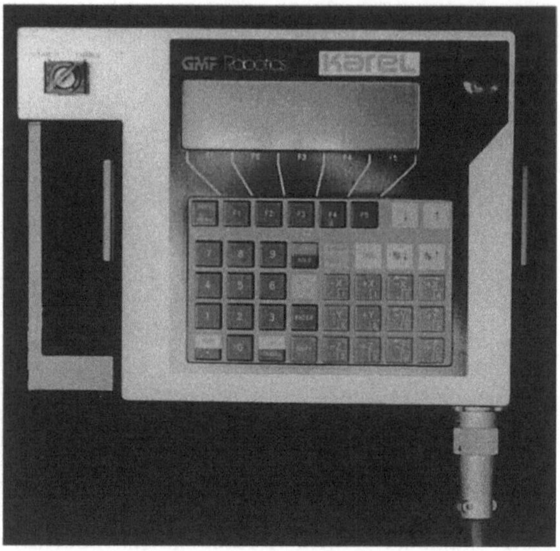

Figure 4.23. Teach pendant of the IR "Karel."

of visual control gives better results, but the best results were obtained, naturally, when the HO used both visual and force feedback channels (Fig. 4.22).

The main advantage of this approach to telerobot sensing seems to be its full independence from characteristics of the force feedback channel from a robot mechanical construction and its control system. Thus, this technique can always be used without essential alterations of the control system.

4.3. On-Off Control

These very wide-spread techniques are described in literature in details (e.g., (Parsons, 1992)). As example of these IDs, the teach pendant of the IR "Karel" is shown in Figure 4.23. Concerning all operation stages with this teach pendant, we refer to the paper (Gladovski *et al.*, 1992).

Briefly, while controlling a robot by using this method, the HO presses in turn pushbuttons corresponding to different DOFs of a robot. As a result of the alternative motions of these drives, a robot endpoint moves in the direction of a target point. To obtain the convergent character of this movement, there must be an additional control element; that is, some kind of regulator which is always situated on the teach pendant to vary the speed of the movements initialized by pushbuttons. While the HO teaches the robot, he/she tunes this parameter after a series of steps toward the target point. This process is continued until the final operations the HO carries out with very low velocities of robot motion corresponding to fine targeting. As this takes place, while drive velocities are decreased, robot arm positioning accuracy is increased. This means one can obtain practically any accuracy level and the method is notable for its simplicity; however, it demands control times that are too lengthly.

Among the positive qualities are:

1) The extraordinary low cost of the equipment for this method. The simple constructions of the teach pendants is the principal merit of this method. This is combined with the absolute absence of computational means because problems to solve inverse kinematics, for example, are left to the HO to solve.

2) The high reliability of these means carried out from simple elements.

3) The low qualification of required personnel performing the robot programming.

4.4. Master-Slave Control

MSSs were the first type of manipulation devices in practical use and they are correspondingly well-described in the literature (e.g., (Thring, 1983; Todd, 1986; Vertut and Coiffet, 1985)). We are therefore going to mainly dwell on this type of control method, the problem sites that arise while using it, and some of the means to advance its efficacy.

MSS can be used to control a robot's main kinematic chain; that is, its arm; it is also applied to control a robot hand. In the following sections, we will describe a conventional group of the former devices, and a special type of such devices - a data glove; that is, a sensing glove that implements the master-slave control for the robot hand.

4.4.1. STRUCTURE AND FEATURES OF MSS

We should recall that this method is used when an EO geometry is similar at a certain scale to an ID geometry. The MSS control mode is described in literature as usually applied to anthropomorphic manipulators (Fig. 4.24); however, this control version is, in fact, only a particular case of the method. For example, at Kazakh State University we developed a master-slave control for an acting model of the closed planar manipulator (Djoldasbekov *et al.*, 1983) of the 3rd class (according to the classification of Assur-Artobolevskii) with three frame-situated driving motors (Fig. 4.25). ID joints A_i, B_i, C_i were connected here (as in the typical scheme shown in Fig. 4.24) to similar EO joints A_e, B_e, C_e by servosystems. The HO controls this manipulator by means of a handle (H), which had an additional pushbutton for the on-off control of a gripper vertical shift of the spatial manipulator (Fig. 4.26).

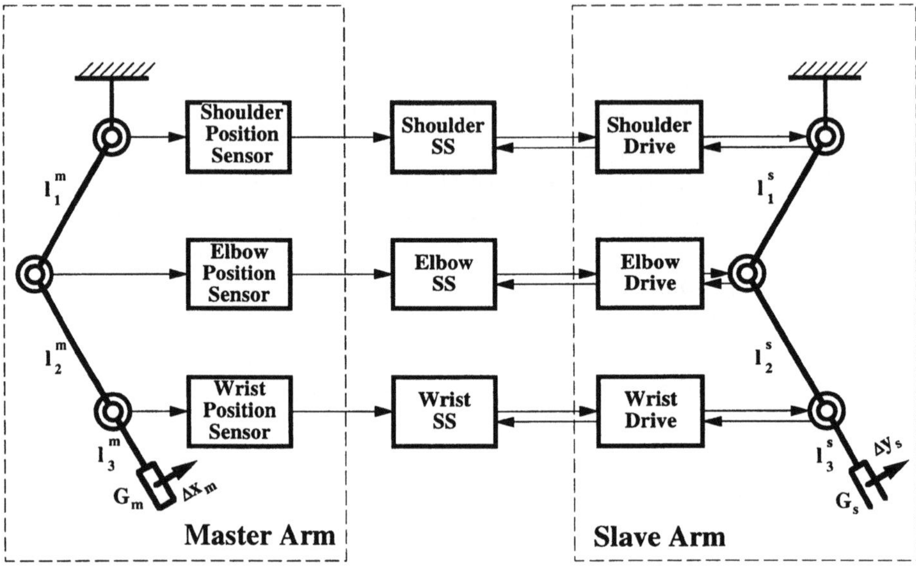

Figure 4.24. Layout example of a MSS. Designation: SS is a servosystem.

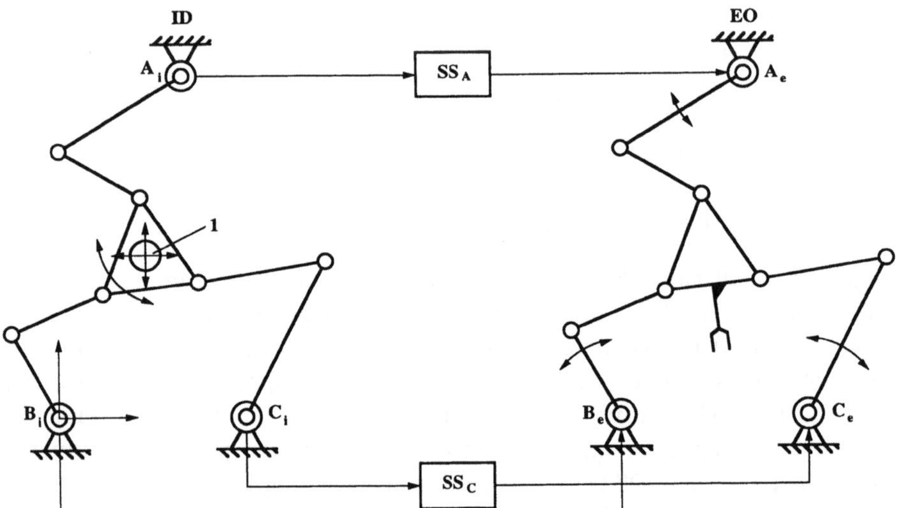

Figure 4.25. Layout of the master-slave control of the closed slave manipulator. SS_A, SS_B, SS_C are servosystems.

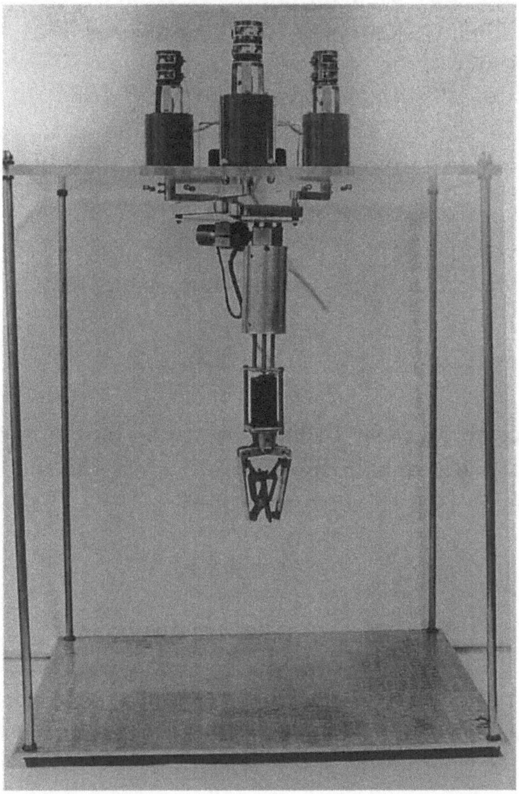

Figure 4.26. Spatial manipulator based on the planar closed mechanism controlled according to the scheme of Figure 4.25.

We pass now to our principal subject; e.g., to the problem of control quality.

4.4.2. QUALITY CHARACTERISTICS OF MSS

Concerning the control rate of these systems, the HO requires no mental calculations to carry out control motions. His/her function is only to move the control handle endpoint in the direction the slave gripper must move, through a distance scaled by the size ratio. At the same time, the inverse kinematics is solved by the master mechanism where the shift signal is distributed on the master axes.

Problems arise because of complexity (i.e. inertialness and nonlinearity) of this electromechanical system controlled by the HO. Such a system is sometimes hard work to move, and, therefore, it is known that the control time in MSSs is greatly increased because of this. Task completion time decrease is possible only by using methods of force reflection from the slave to the master. This creates just the bilateral systems to which we have previously referred.

Let us consider a problem of accuracy in MSSs. In using the system example shown in Figure 4.24, we see that shifts Δx_m of master endpoint G_m and Δx_s of slave endpoint G_s are bound through a certain scale ratio k:

$$\Delta y_s = k\Delta x_m, \tag{4.1}$$

and this ratio depends directly on a relationship of master and slave link lengths

$$k = \frac{l_1^s}{l_1^m} = \frac{l_2^s}{l_2^m} = \frac{l_3^s}{l_3^m}. \tag{4.2}$$

This equality has to be meet to keep complete mnemonicability of control.

Then, by setting a certain resolution Δx_i^* of the manipulator input we can determine from (4.1) a slave resolution

$$\Delta y_s^* = k\Delta x_i^*. \tag{4.3}$$

The value Δx_i^* depends on a master resolution Δx_m^* and HO resolution Δx_H^*.

Let the two random variables be independent and normally distributed. It is known that their limit total value may be determined according to

$$\Delta x_i^* = \sqrt{\Delta x_H^{*2} + \Delta x_m^{*2}}. \tag{4.4}$$

Therefore, this value is greater then each of its components individually.

Such a complicated mechanical system as the master arm has usually a very low resolution and the HO resolution is also not high; therefore the entire system input resolution (4.4) is, in practice, not a very high value. Concerning the scale ratio value in (4.2), there are, principally, two practical cases. In the nuclear industry, for example, there is an opportunity to carry out master and slave arms for work in hot cells of the same size. In this case, the scale ratio (4.3) $k = 1$, and this ensures the maximum accuracy that may be achieved in MSSs.

Another state is for MSSs that are used for IR teaching and to control manipulators that work in extreme and hazardous environments. For all these applications, the deciding factor is an availability of certain overall dimensional restrictions on control handle transferences. These restrictions concern both master dimensions and overall dimensions of control shifts of the HO arm. In this case, the master is usually of small dimensions. Such a master is also called a "replica" (Jacobus *et al.*, 1992). The scale ratio is then $k << 1$ (e.g., 1:6.2 (Wilt *et al.*, 1977)), and it follows from

(4.3), the MSS resolution is considerably decreased (when characteristics of servosystems are constant).

In fact, system resolution is not only a function of the mechanical ratio, but also depends on amplification factors in control systems between master and slave arms (Fig. 4.24). However, it is necessary to remember that, while changing these factors, a disconcordance of these arm postures occurs which decreases control quality, as was shown by results presented in Subsection 3.3.1.

These are general features of master-slave control application in these fields, and, as result, it is used frequently to solve manipulation problems that are known to be somewhat inaccurate. In the industrial robotics field, these are, for instance, robotic loading manipulators of high carrying capacity and IRs whose task is painting.

4.4.3. FLEXIBLE MASTER TO CONTROL A SLAVE ARM

In MSSs, the slightest alteration in the construction of a slave arm demands a corresponding alteration in the construction of a master, or additional kinematic transformations (Jansen *et al.*, 1992), because of control quality aggravation. In addition, it is common knowledge that a master-slave regime application for multi-joint redundant manipulators is impossible (Asano, 1984). At the same time, such a possibility is very attractive because of difficulties in this case with computer controlled algorithm implementation in real time.

It has been suggested removing some of the deficiencies of conventional MSSs through the application of a flexible element as the basis of the master manipulator (Slutski, 1986; Slutski, 1995). Based on the flexible element, the master takes up a certain pose while it is being loaded. An innovative control concept is that, as this takes place, a slave is inscribed within a given flexible line of the master on a certain scale. This line is approximated by a polygonal line that is geometrically similar to the slave configuration.

The principle behind a master based on a flexible working element is illustrated in Figure 4.27. The flexible master (FM) is shown here in the form of a rigidly clamped flat spring (S), on the free end of which the control handle (R) is fixed. The sensors $(S_i), (i = 1, ..., n)$ are placed on the spring (in this example, they are strain gauges). The spring-bending deformations are used here to set swing angles, while torsion deformations may be used for setting rotation angles of the corresponding slave joints.

In Figure 4.28, we show an example of how such a system can control a planar manipulator with two DOFs. The slave arm has two drives, (D_1, D_2), in accordance with which pairs of sensors (S_1, S_2) (strain gauges) are placed on the flat spring. In this case, we are dealing with gauges that

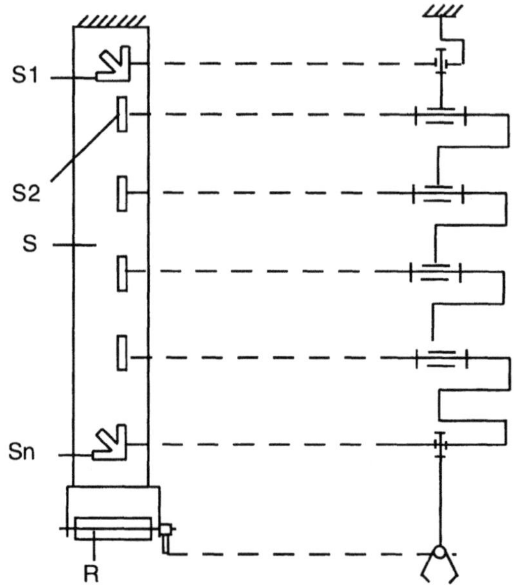

Figure 4.27. Example of organization of a master based on a flexible working element. Designations: R is a master handle, S is a flat spring, and $S_1 - S_n$ are strain gauges.

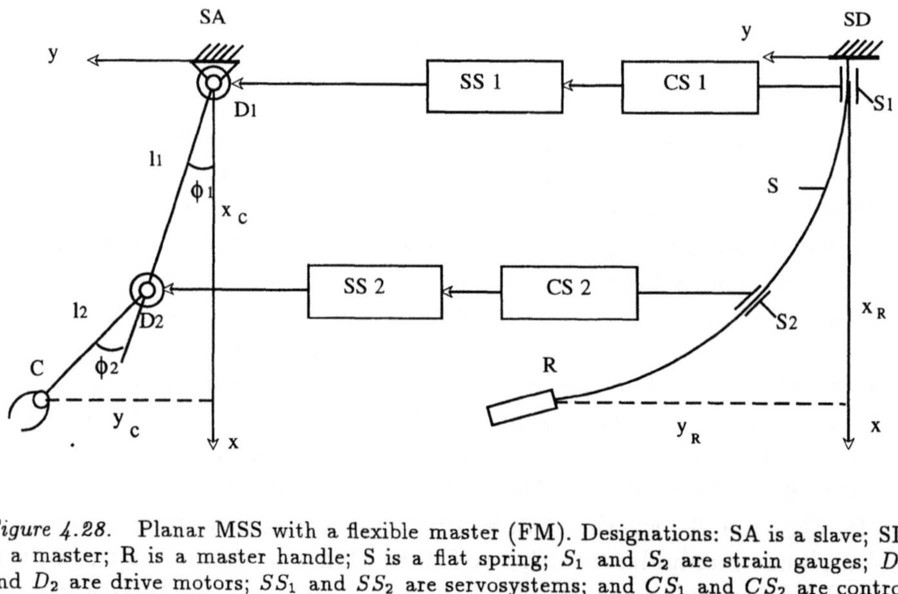

Figure 4.28. Planar MSS with a flexible master (FM). Designations: SA is a slave; SD is a master; R is a master handle; S is a flat spring; S_1 and S_2 are strain gauges; D_1 and D_2 are drive motors; SS_1 and SS_2 are servosystems; and CS_1 and CS_2 are control measuring schemes.

are placed on both sides of the spring for an increase of measurement reliability.

Controlling this master, the HO manipulates it with the help of handle (R), setting the necessary configuration of the spring. Signals from the strain gauges then enter the measuring bridge schemes $(CS_i)(i = 1, 2)$, where they are also amplified and transferred to the corresponding servosystems $(SS_i)(i = 1, 2)$. The control system principle is that relative angles of turns in the slave joints must be proportional to spring strains measured at the points corresponding to these joints. Gains of the units CS_i are chosen so that the obtained configuration of the slave is inscribed within a given flexible line of the master.

The number of sensors placed on the spring may be more than a number of DOF of a concrete manipulator. It is of fundamental importance that there are redundant sensors, and that a certain necessary number of them are connected to the control system when needed. As a rule, this number is equal to the DOF number of the manipulator, but other solutions may also work and will be listed below.

Under spring-bending, signals that are proportional to the values of the relative deformations are taken from the gauges. These signals are intensified in the control system to form control signals

$$\phi_i = \tau_i \varepsilon_i, \; i = 1...n \; , \tag{4.5}$$

where ϕ_i is the turn angle of the i-th link relative to the previous link; τ_i is the gain in the i-th channel of the control system; ε_i is the relative deformation in the i-th point of the flat spring, which corresponds to the i-th joint of the slave; and, n is the number of sensors on the flexible element - equal in this case to the number of controlled DOFs of the slave.

Control system gain values for each concrete pose of the master and the slave can be determined, because the positions of the slave gripper (C) and the master handle (R) (Fig. 4.28) are coincident in the working plane of the Cartesian space

$$x_C = \mu x_R, y_C = \mu y_R \; ,$$

where μ is a scale coefficient.

The angles ϕ_i for the expression (4.5) are determined by solving inverse kinematics for a corresponding manipulator.

The master point R coordinates depend on values of a force $\vec{P}$ and a moment M applied to the flat spring

$$x_R = x_R(\vec{P}, M), y_R = y_R(\vec{P}, M).$$

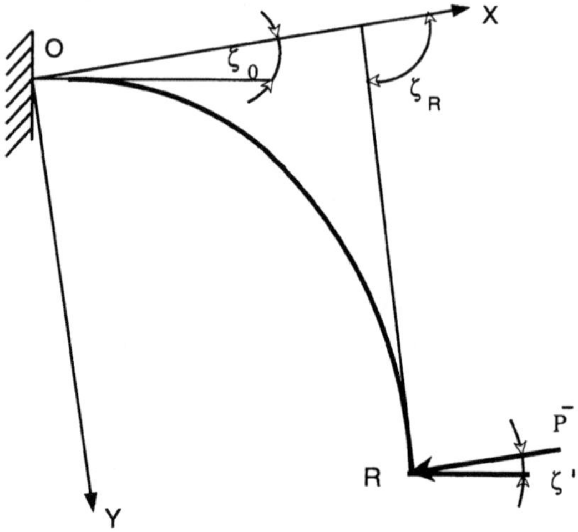

Figure 4.29. Calculation scheme of the FM.

Flexible Element Calculation

It is known that when the origin of coordinates is at a point of spring restraint, and the abscissa axis is directed along the action line of applied force (Fig. 4.29), the elastic line of a flat spring is described for noninflected shapes by equations (Ponomariov and Andreeva, 1980)

$$\frac{x_i - x_0}{l} = \frac{2}{\kappa\beta}[E(\kappa, \psi_i) - E(\kappa, \psi_0)] - \left(\frac{2}{\kappa^2} - 1\right)\frac{S}{l},$$

$$\frac{y_i - y_0}{l} = \frac{2}{\kappa\beta}[K(\kappa, \psi_0) - K(\kappa, \psi_i)], \tag{4.6}$$

where x_0, y_0 are coordinates of the point of spring restraint on the base; l is the spring length; κ is an integration constant that depends on the load, applied to the spring, and its line of action; $E(\kappa, \psi_i)$ is the elliptic integral of the second kind; $K(\kappa, \psi_i)$ is the elliptic function; S is the spring segment length counted from the origin of coordinates; $\psi_i = \frac{\zeta_i}{2}$, ζ_i is the slope angle of the tangent with respect to the spring's elastic line in the i-th point; and, $\beta = l\sqrt{PB^{-1}}$ is the dimensionless load parameter with P value of the force acting to the spring and B spring flexural rigidity.

For noninflected shapes of the spring's elastic line, the following equations are added to (4.6):

$$\beta = \kappa[F(\kappa, \psi_R) - F(\kappa, \psi_0)],$$

$$F(\kappa, \psi_i) = [F(\kappa, \psi_R) - F(\kappa, \psi_0)]\frac{S}{l} + F(\kappa, \psi_0) ,$$

where $F(\kappa, \psi_i)$ is the elliptic integral of the first kind.

The bending moment value at the spring's i-th point is then determined (because $\zeta_0 = \zeta'$) from the expression

$$M_i = \frac{2K(\kappa, \psi_i)}{\kappa\beta} Pl + \frac{B}{R_c} , \tag{4.7}$$

where R_c is the initial curvature radius of the spring's elastic line (generally, $R_c = \infty$).

If we know the bending moment value, it is possible to determine a relative deformation value by using the formula

$$\varepsilon_i = \frac{M_i h}{2B} , \tag{4.8}$$

where h is spring thickness. In this case, the deflection of a flexible spring made of a thin strip can be of the same order of magnitude as the strip length. Such large deflections are accompanied by considerable change in the curvature of the spring axis. Thus, the spring works at the field of great displacements but small elastic deformations, because of the small h at the expression (4.8) (Ponomariov and Andreeva, 1980).

The parameter h may be chosen on the basis of the following considerations:

a) If the spring thickness is sufficiently small, spring deformations may remain small and elastic:

$$\varepsilon_{max} = \frac{M_{max} h}{2B} < \varepsilon_l , \tag{4.9}$$

where ε_{max} is the greatest deformation in the bending spring; ε_l is the deformation corresponding to a material's elasticity limit; and, M_{max} is the maximum value of a load.

b) On the other hand, the maximum gain value τ_{max} must not exceed a certain value τ_l, because interferences will then have too much influence on control signals:

$$\tau_{max} = \frac{\phi_{max}}{\varepsilon_{max}} = \frac{2\phi_{max} B}{h M_{max}} < \tau_l . \tag{4.10}$$

By solving these two inequalities (4.9) and (4.10) relative to h, we find a range of possible spring thickness

$$\frac{2B\phi_{max}}{M_{max}\tau_l} < h < \frac{2B\varepsilon_l}{M_{max}} . \tag{4.11}$$

When the desired values of the turn angles ϕ_i of the slave arm links and the relative deformation values in the corresponding points of the master have been determined, the values of gains τ_1, τ_2 may be found from equations (4.5), (4.7), and (4.8).

In solving this problem, we used the following approach to achieve simulation simplicity. We set a bending spring's elastic line shape; we then determined load values that corresponded to the given form of the spring, and the relative deformation values at the points in which we were interested. After this, the slave pose, which corresponds to the shape of the bending spring's flexible line, was determined, and the gains τ_1 and τ_2 were then calculated.

A model of such a system, consisting of a two-link slave and a master based on a flexible element with strain gauges, was obtained by using the described approach. Manipulator parameters used during simulation were: $l_1 = 0.45$ m, and $l_2 = 0.35$ m. The spring parameters were: the modulus of elasticity $E = 2.1 \cdot 10^5$ mPa; cross-section sizes $b = 3 \cdot 10^{-2}$ m; $h = 0.1 \cdot 10^{-2}$ m (these satisfy the inequality (4.11)); and length 0.8 m.

In accordance with the described method, the gains $\tau_i, i = 1, 2$ were computer calculated for a number ($m = 60$) of different master poses. The results obtained indicated that the gains' values are changed at a certain interval depending on load character. This makes it possible to determine their average value for each control channel

$$\bar{\tau}_i = \frac{1}{m} \sum_{j=1}^{m} \tau_{ij} ,$$

where $\bar{\tau}_i$ is the average gain value for i-th channel; τ_{ij} is i-th channel gain for j-th configuration of the bending spring; and, m is the total number of configurations.

The following values for the average gains were obtained:

$$\bar{\tau}_1 = 20651, \bar{\tau}_2 = 57401 . \tag{4.12}$$

We take these values as actual gains. When using them, system positioning accuracy is decreased; that is, there is an increase in the difference between coordinates of the slave EE and the master handle. Let us consider this deflection value. For this, we may determine differences between the theoretical and the actual (obtained by using the average gains (4.12)) coordinates of the manipulator EE for each point of the workspace. However, the mnemonicability coefficient (see Chapter 3) gives us a more weighted estimate of system accuracy at a studied point.

Mnemonicability of Control System with FM
Let us set shift vectors of the master endpoint $\vec{R}_i = \vec{R}(x_{Ri} - x_{R0}, y_{Ri} - y_{R0})$ to different directions with modulus equal to one. The slave endpoint is then moved in definite directions which are characterized by vectors $\vec{C}_i = \vec{C}(x_{Ci} - x_{C0}, y_{Ci} - y_{C0})$. We determine the coordinates of points (x_{Ci}, y_{Ci}) by using the above-described method. For the mnemonicability coefficient calculation, we determine the angle between the shift vectors $\vec{R}_i$ and $\vec{C}_i$ of master and slave endpoints

$$\cos v_i = \frac{(x_{Ri} - x_{R0})(x_{Ci} - x_{C0}) + (y_{Ri} - y_{R0})(y_{Ci} - y_{C0})}{\sqrt{(x_{Ri} - x_{R0})^2 + (y_{Ri} - y_{R0})^2}\sqrt{(x_{Ci} - x_{C0})^2 + (y_{Ci} - y_{C0})^2}} .$$

By using p shift vectors of the master endpoint to select different directions, we determine the system's mnemonicability coefficient for this point (x_{Ri}, y_{Ri}) within the workspace:

$$K_M = \frac{1}{p}\sum_{i=1}^{p} \cos v_i . \tag{4.13}$$

Based on the use of the average gains (4.12), this algorithm was computer-calculated for eight points within the working plane and for eight different directions (the number $p = 8$ in expression (4.13)). The obtained results are combined and presented in Table 4.1 and indicate that the mnemonicability coefficient is sufficiently big. Its average value for the discussed points is

# pose	1	2	3	4	5	6	7	8
x	-0.15	-0.15	-0.1	-0.1	-0.05	-0.05	0	0.05
y	0.15	0.2	0.2	0.25	0.25	0.3	0.3	-0.3
K_M	0.7383	0.9810	0.8140	0.8965	0.7077	0.7409	0.9474	0.7493

TABLE 4.1. Values of mnemonicability coefficients for different points of the working plane.

$K_M = 0.79$ (Djamalov and Slutski, 1986). It is known that a manipulation system efficiency under this coefficient value remains satisfactory enough (Chapter 3).

By using the strain gauges, it is possible to very easily organize the spring's inflected shapes to control multi-link manipulators. For instance, a carriage (C) with an additional handle may be placed on the spring (Fig. 4.30). The HO controls the additional handle (AH) by using the second hand while his/her first hand controls the master's main handle (H).

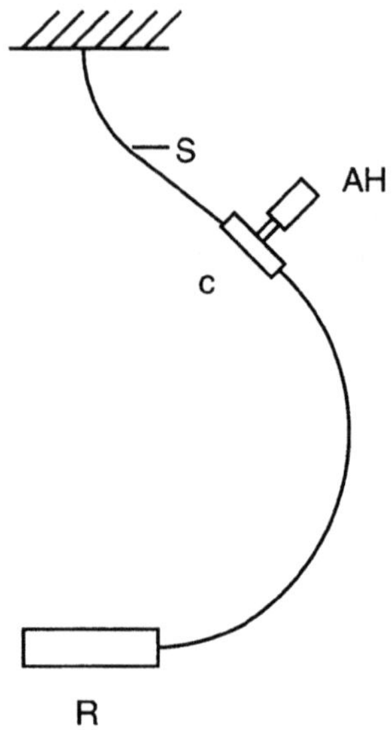

Figure 4.30. Layout of the FM with additional handle. Designations: S is a flat spring; C is a carriage, R is a main handle, and AH is an additional handle.

Master Based on Potentiometer Gauges

It is known that a strain gauge only maintains its capacity to work within a certain range of values of relative deformations in flexible element. As a result, there must be some restrictions on flexible element cross-sizes and mechanical properties. Therefore, it is worth considering another type of FM that uses potentiometer gauges and has no such restrictions.

Such a device is based on the principle that lines that are perpendicular to a flexible element's neutral axis before deformation also remain perpendicular after deformation. The next two sections of the flexible element are turned at a certain angle (Fig. 4.31). Thus, in the figure, posts (2), placed on flexible element (1), are perpendicular to its surface at all times. The potentiometers (3) are attached to the posts (2). Each potentiometer axis is spring-mounted and is connected by the flexible traction (4) to the next post. During flexible element deformation, the next two posts are turned relative to one another at a certain angle ω_i and then the lengths of the flexible tractions are changed. In this case, it is possible to directly approximate the form of the flexible element by means of a polygonal line; e.g., on the basis of the flexible tractions. Alterations in the lengths of the flexible tractions l_i are measured by the potentiometer gauges.

Two versions of the posts' placement on the FM may be considered. In the first version, the posts are placed on one side of the flexible element

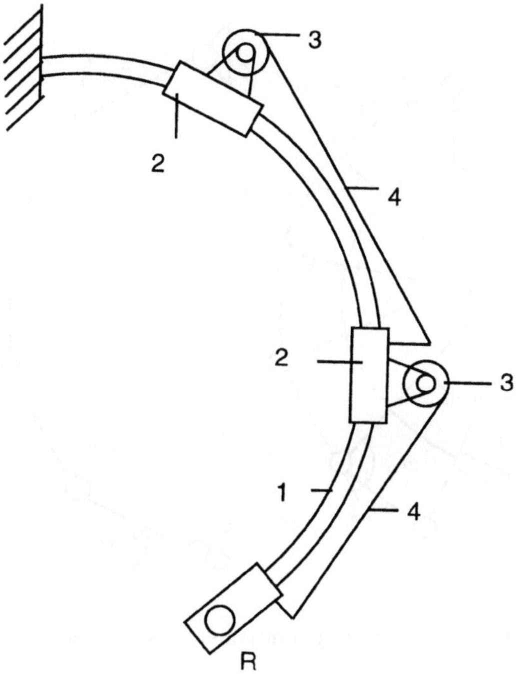

Figure 4.31. The FM with potentiometer gauges. Designations: 1 is a flexible element, 2 are posts, 3 are potentiometer gauges, and 4 are flexible tractions.

(Fig. 4.32). In this case, a dependence

$$l_i = 2\left(\frac{l_{Di}}{\omega_i} \pm H\right)\sin\frac{\omega_i}{2}, i = 1, ..., n \tag{4.14}$$

exists between the values l_i and ω_i on condition that the form of that section of the flexible element between the next posts is closely applied to the circle arc. The following designations are used in formula (4.14): l_{Di} is the length of flexible element section between the next two posts; H is the height of the posts; and n is the number of main points (e.g., posts). A sign before value H is determined by the direction of flexible element deformation.

In the second version, the posts with position gauges are disposed symmetrically and bilaterally along both sides of the flexible element (Fig. 4.33). In this case, a dependence between values l_i and ω_i will look like

$$\omega_i = \arccos\left(1 - \frac{1}{2}\left(\frac{\Delta l_{12}^i}{2H}\right)^2\right), \tag{4.15}$$

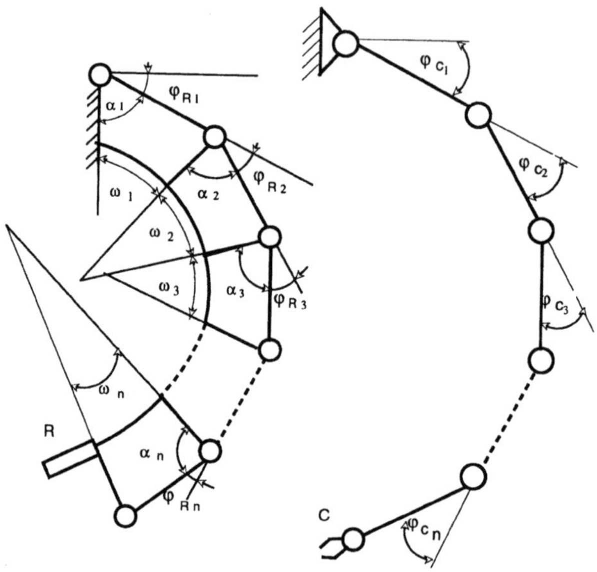

Figure 4.32. Placement of the potentiometer gauges on one side of the FM.

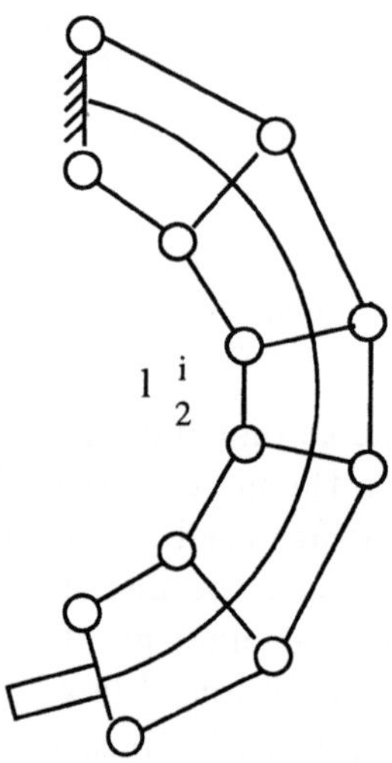

Figure 4.33. Placement of the potentiometer gauges on both sides of the FM.

where Δl_{12}^i is a difference between the lengths l_1^i and l_2^i of the flexible tractions placed on the opposite sides of the FM.

For both these versions, the relative angles ϕ_{Ri} of the turn of the flexible tractions are determined by the following expression:

$$\phi_{Ri} = \frac{\omega_{i-1} + \omega_i}{2} , \tag{4.16}$$

where the values ω_i, ω_{i-1}, $i = 1, ..., n$ are obtained from expressions (4.14) or (4.15).

It is obvious that precision of approximation, and, consequently, control precision, depends on the number of the main points; that is, on the quantity of the posts with sensors placed on the FM. However, the posts may be positioned at a certain distance l_D along the flexible element (Fig. 4.33), which is characterized by the relationship

$$2H \arccos \frac{R_m}{R_m + H} < l_D < 2R_m \arccos \frac{R_m}{R_m + H} .$$

The notation R_m denotes a permissible radius of curvature, to which it is possible to bend the spring, and it does not transfer a limit of elasticity.

To simulate or to design this type of system, we have to use the calculation methods described above. However, to control the multi-link slave by using this master and to inscribe it within some shape of flexible element bend, it is possible to implement simple dependences,

$$\phi_{Ci} = \phi_{Ri}, i = 1, ..., n , \tag{4.17}$$

(Fig. 4.32) in the control system.

It can be determined that, when implementing (4.17), the control precision is reduced when flexible element deformation increases. Quantitative evaluation of control quality was made with the help of the mnemonicability coefficient by using mathematical models of different manipulation systems.

Thus, our example system, which consists of a two-link slave and a master based on an flexible element with two gauges, has an average mnemonicability coefficient $K_M = 0.76$.

A system consisting of a two-link slave and a master with eight sensors, five of which work on the first link and the rest on the second link, has an average mnemonicability coefficient of $K_M = 0.95$.

A system consisting of an eight-link manipulator and a master with eight sensors, has an average mnemonicability coefficient of $K_M = 0.98$.

It therefore turns out that the greater the quantity, both of the posts with sensors and of the DOF of the slave, the greater the mnemonicability level for the developed control method.

An example of such a device for a spatial case is presented in Figures 4.34 and 4.35.

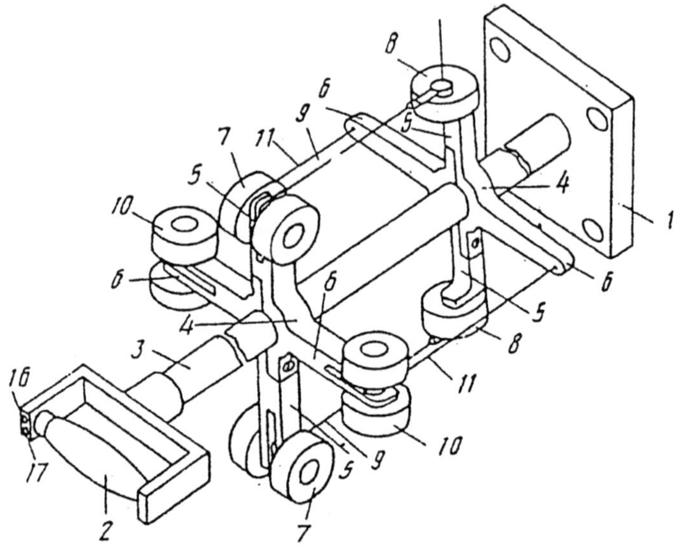

Figure 4.34. Setup of the FM to control spatial manipulators. The principal designations are: 1 is a base; 2 is a handle; 3 is a flexible element; 4 are platforms; 7, 8, and 10 are potentiometer gauges.

This device contains a flexible element (3) of the circular section that connects the base (1) and the handle (2). Platforms (4) with posts (5) and (6) are positioned on the element (3). The potentiometer gauges (7), (8), and (10) are placed on these posts. Deformation of the flexible element (3) changes the pull of the flexible tractions (9) and (11). The axes (12) and (13) of the potentiometers (7) and (10) may be backed by the springs (14) and provided by the brakes (15) (Fig. 4.35). The brakes are connected to load sensors of a slave through a control system.

This master (Djoldasbekov *et al.*, 1990) permits control by the HO of the spatial multi-link manipulator. By means of brakes on the master posts, the device allows, in principle, the provision of force feedback to the HO. When loads arise on the slave, the brakes reflect them to the master by preventing flexible element bending. The picture of this master (without sensing elements) is shown in Figure 4.36.

4.4.4. DATA GLOVE

As was already noted above, data gloves serve to implement master-slave control of telerobot hands. A number of varieties of these devices is described in Burdea and Coiffet's book (1994). The data glove may be used

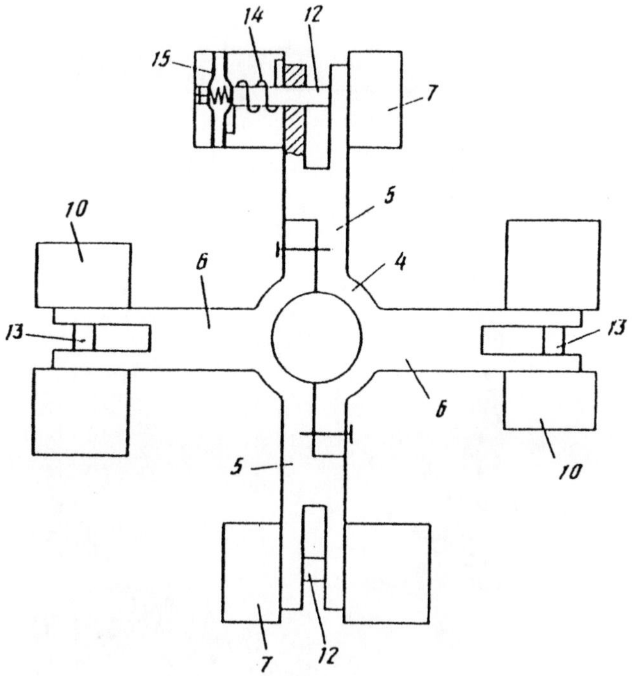

Figure 4.35. Placement of sensors and brakes on the posts of the spatial FM. The principal designations are: 5 and 6 are posts; 7 and 10 are potentiometers; 15 is a brake.

both to control real telerobots (Fig. 4.37) and to control virtual reality objects (Fig. 4.38).

It is necessary to remember that the data glove sets only signals to control a robot hand; therefore, to control the whole robot, the glove has to be combined with a conventional HC as described in (Bejczy, 1994; Burdea, 1989).

4.5. Semiautomatic Control

Semiautomatic control (SAC) was created as an alternative to remote control methods (described above), which are not able to ensure the required characteristics of telecontrol quality. The fundamental feature of SAC is that ID kinematics are dissimilar to EO kinematics. The second feature follows from the first, and, in this systems structure, is a special computing device which performs coordinate transformations. To carry out all necessary kinematic transformations in the computing device in real-time, input signals from ID are sampled, for example, by 60,...,200 times/sec (Jacobus *et al.*, 1992).

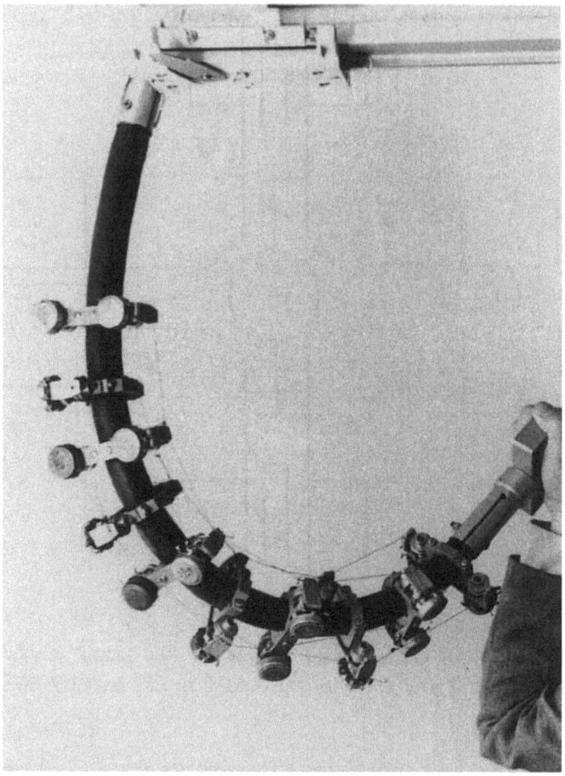

Figure 4.36. Picture of the spatial FM.

A principal idea of SAC is in setting a specific vector by using a HC, and the EO endpoint has to be moved according to this vector. The vector dimensions determine the HC structure. The most widespread case consists of forming 3D vectors to control a main kinematic chain of a spatial manipulator. This is based on the fact that problems such as IR accuracy and fast-operation are primarily solved while moving the main kinematic chain. As for local (orientation) motions, they do not essentially influence the level of resultant indices of accuracy and fast-operation because of their small amplitudes. Moreover, for IRs, for example, local motions are frequently actuated with the help of opened pneumatic or hydraulic drives that do not permit the application of the SAC approach.

On the other hand, for multi-DOF telerobots, the number of set signals must be more than three. For instance, in non-industrial applications, control of orientation DOFs, as well as control of movements of an additional

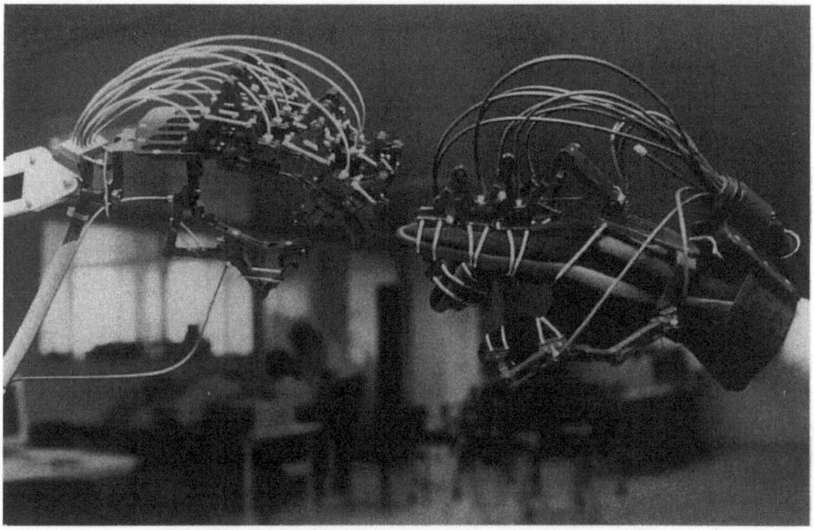

Figure 4.37. The data glove to control multi-finger robot hand. Courtesy of P. Coiffet, Laboratoire de Robotique de Paris.

Figure 4.38. The data glove to control a robot hand in a virtual reality environment. Courtesy of P. Coiffet, Laboratoire de Robotique de Paris.

small arm attached to a large manipulator (see, e.g., the manipulator of Fig. 2.1), is also carried out by continuous motions of a HC because of the high demands of the quality of the control processes. There are two possible

Figure 4.39. Two hand controllers to control the manipulator for forestry operations. Courtesy of A. Goldenberg, University of Toronto, Canada.

solutions of this problem. The first is associated with the creation of multi-axis HCs to control one multi-DOFs manipulator. This approach will be described below in this Section. It should be noted, however, that control of such a multi-component HC is awkward for the HO and is fraught with his/her possible mistakes during the control process.

The second solution uses IDs (usually, two) to control different groups of telerobot DOFs. As this takes place, the HO controls two HCs simultaneously using both hands. As a rule, control is effected by means of the first HC by robot transportation DOFs, with the help of the second HC by motions of robot orientation DOFs. An example of such a control station with two HCs is shown in Figure 4.39 for a manipulator for forestry operations (Fig. 2.5).

However, this approach is not always completely successful. First, it is difficult for the HO to co-ordinate and simultaneously carry out different motions of both of his/her arms. Second, there are some industrial circumstances when the HO does not have the possibility to make use both of his/her arms simultaneously. Therefore, a highly expedient version of an ID is a 3-axis HC which control orientation motions by means of different switchers and pushbuttons placed on the handgrip. If, while controlling a telerobot, however, there is a necessity to realize smooth motions of the orientation DOFs, another option may be used, which consists of the use of the same HC to set signals to control both transportation and orientation motions. As this takes place, the HO implements, first, one control mode, and, after that, another one. He/she carries this transition out by means of a special selector placed on the handle. It is clear that control process time is increased in this case. There is also a necessity here to incorporate special control devices that fix manipulator generalized coordinates in joints not being controlled at a given moment.

4.5.1. ALGORITHMS AND STRUCTURES OF THE SAC SYSTEMS

A maximum composition of functions performed by a system computing device is as follows:
1) kinematic transformation of signals from the ID;
2) calculations to implement proper control modes of the system;
3) kinematic transformations of signals that enter EO drives;
4) processing and transformations of sensing signals that enter from force sensors placed on the EO.
There are also a number of other functions that are typical of telerobot control systems, but these are not specific functions of the SAC systems. They are:
5) an adjustment of dynamic characteristics of robot servodrives;
6) a compensation of external forces and moments, acting to a manipulator, and static moments of gravity forces of manipulator links;
7) compensation of dynamic cross-coupling manipulator DOFs.

From this set of functions, from one to all seven may be implemented within a concrete control system structure. As this takes place, almost all of them are typical functions, calculated according to known, previously developed algorithms. Note that some of the functions may be absent in an SAC system as dictated by the characteristics of the ID and EO used in the system. For instance, there is no need for functions 1, 3, and 7 when the SAC of the Cartesian manipulator is completed by the Cartesian ID. The 4th function, which consists of feeding back forces from the EO, is also not used, for example, in the IR teaching systems.

However, only the second of the listed functions is always characteristic of SAC systems, and its implementation is not trivial. Thus, a consider-

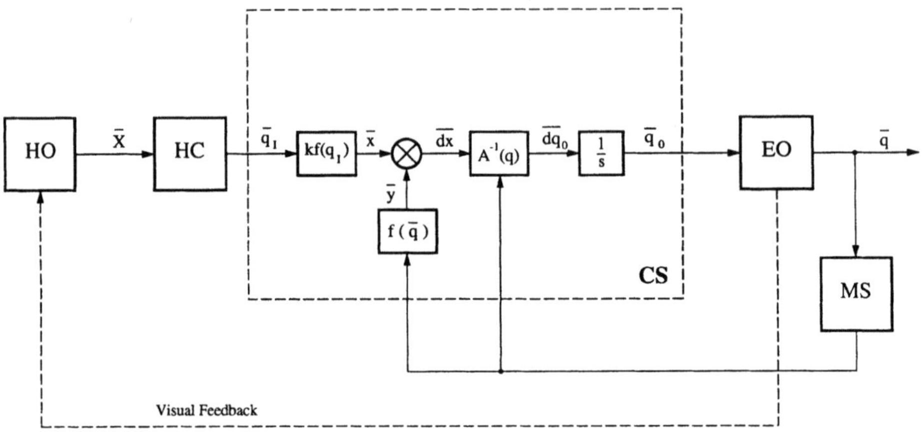

Figure 4.40. Block diagram of a position control system (CS is a control system and MS is a measurement sensor system).

able part of this work is devoted to the development of the function of SAC systems. In turn, this function is defined by a chosen control mode of teleoperations.

Position Control

A layout as shown in Figure 4.40 is usually used for the SAC of position type systems. In this case, the HO controls the HC by setting Cartesian coordinates $\vec{X}$ of its endpoint. The HC's generalized coordinates $\vec{q}_I$ then enter a CS, and control signals $\vec{q}_O$ from this block enter robot EO servo-drives. EO output signals are measured by a MS. The CS is implemented as a computing device performing specific operations. The device first transforms values of generalized HC coordinates $\vec{q}_I$ into Cartesian coordinates $\vec{x}$, which are associated with a HC basis. This transformation is carried out with some scaling determined by a constant k (Fig. 4.40).

To be precise, this coordinate transformation may sometimes be absent if values $\vec{x}$ are set directly by the HC. Likewise, a block that transforms values of generalized manipulator coordinates $\vec{q}$ to a spatial position $\vec{y}$ of its gripper may be absent in the system structure. Elemental differences of coordinates $\delta\vec{q}_0$ in manipulator joints are calculated by solving inverse kinematics for an arm mechanism

$$\delta\vec{q}_O = A^{-1}(\vec{q})\delta\vec{x},\ \delta\vec{x} = \vec{x} - \vec{y}, \tag{4.18}$$

where the Jacobian matrix A ties generalized velocities $\dot{\vec{q}}_0$ in manipulator joints and a gripper velocity $\dot{\vec{x}}$ in a Cartesian base frame:

$$\dot{\vec{x}} = A\dot{\vec{q}}_0. \tag{4.19}$$

This approach to solve a position control problem is called "the resolved motion rate control," and it is often used in general robotics (Klafter *et al.*, 1989).

Principally, there is another type of this problem solution. In fact, there is a possibility to solve inverse kinematics, not for velocities (4.18), but directly for positions. For equations that connect desired Cartesian coordinates $\vec{x}$ of manipulator endpoint with values of its generalized coordinates $\vec{q}_O$, we have:

$$\vec{x} = F(\vec{q}_O), \tag{4.20}$$

with the following solution:

$$\vec{q}_O = F^{-1}(\vec{x}). \tag{4.21}$$

This general approach, which uses an inverse kinematics solution, is called "the resolved motion position control" (Klafter *et al.*, 1989).

However, a supplementary complication is that the vector-function F (4.20) is non-linear and represents a system of transcendental expressions. A solution of the vector equation (4.21) is therefore almost always non unique. Its complete solution includes all possible alternative assemblies of a manipulator mechanism. Because of this, an inverse kinematics solution for positions is more difficult than for velocities. The computing device not only has to solve a difficult mathematical problem, but also to extract and to maintain a certain configuration of the manipulator. We would like to refer here to some of the papers contributed by, for instance, Prof. B. Roth for different groups of these problems. One paper on methods for general solutions of inverse kinematics for series and in-parallel chains is by Raghavan and Roth (1995). Other papers (Raghavan and Roth, 1990a; Raghavan and Roth, 1990b) present a method for the series chain configurations. As is known, for the in-parallel manipulator structures (or platform manipulators), inverse kinematics are solved more simply than forward kinematics (in contrast to opened kinematic chains). Therefore, for these mechanisms, the latter problem is essentially an object of investigations and obtained solutions are less general than the series, since they deal with very special cases of specialized structures (Nielsen and Roth, 1996).

This leads to more complicated software for the control system and an increase of computational time - although the system structure, while calculating algorithm (4.21), is somewhat simplified compared with the system related to algorithm (4.18). This is also because, to solve inverse kinematics for velocities, a measurement of generalized coordinates is required.

In any rate, for both described approaches, a basical kinematic relationship between HC input control signals $\vec{X}$ and robot endpoint coordinates $\vec{Y}$ should be performed for the position control

$$\vec{Y}(t) = k\vec{X}(t), \ k = const. \tag{4.22}$$

Thus, when using the semiautomatic position control mode, a character of HO motions corresponds to those in the case of the master-slave control. The difference in this case is only that the corresponding inverse problem is solved, not by means of an arm mechanism, but by using kinematic transformations in the computing device. In SAC systems, there is also the coefficient of transformation k (4.22), which determines a number of system characteristics. Just by tuning this parameter, it is possible to change output indices of the system by moving its working point along the line shown in Figure 1.3; for example, maximum resolution of the system is achieved when the value of k is minimum. On the other hand, low EE velocity and larger ID overall dimensions occur.

Rate Control

This control mode is primarily intended to implement EE transportation motions with a principal demand of high velocity. The essence of the method is that a control vector set on the HC and a velocity vector of the robot EE shift are matched one-to-one. The experience of Kan and Austin (1990) showed that there are difficulties in implementing the rate semiautomatic control for the robot rotational axes. The authors concluded that pitch, yaw, and roll motions become impossible to comprehend after a short excursion from the neutral position, and that the cause is the cross-coupling between axes. Thus, for orientation inputs, it is recommended to use the position control.

While implementing the rate control, the approach usually in the use is that vector $\vec{X}$ is formed directly on the HC, and a necessary vector of robot endpoint velocity $\dot{\vec{Y}}$ is set based on it:

$$\dot{\vec{Y}} = \eta\vec{X}, \ \eta = const.$$

In this case, the control system layout is the same used as the basis of the position control system (Fig. 4.40); as shown in Figure 4.41. Vector $\dot{\vec{x}}$ is obtained after corresponding kinematic transformations are connected to generalized velocities $\dot{\vec{q}}_O$ in the arm joints by the same relationship (4.19).

A principal prescription of the rate control is the realization of fast manipulator motions in its free workspace. These systems obviously do not ensure a high level of positioning accuracy, since a speed criterion is in conflict with an accuracy criterion. This is one basis on which to organize combined control systems.

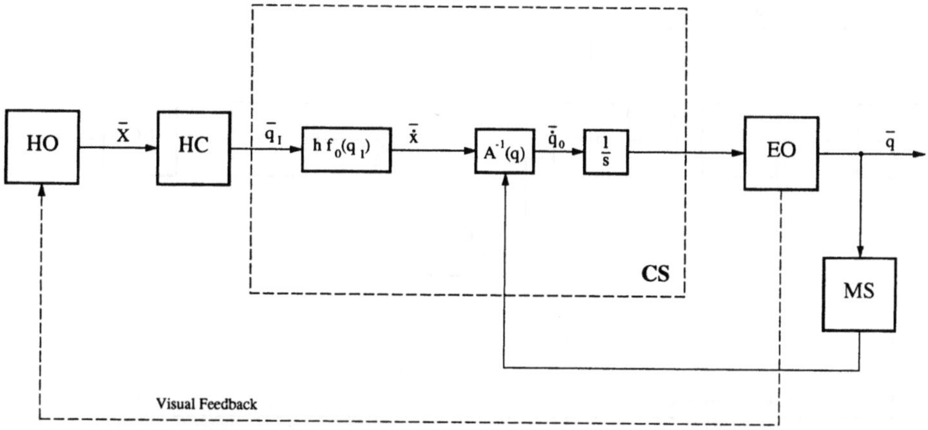

Figure 4.41. Rate control system block diagram. Designations are the same as those in Figure 4.40.

Combined Control

Combined control consists of combining above-described control modes (including force (acceleration) control, if needed) in cases when each of these modes separately does not ensure a required quality of system operations. The necessity of combined control arises when a telerobot has to perform varied operations resulting from different subtasks. A typical task of remote manipulation systems might be having the robot take a certain object, move it to a considerable distance from its initial point, and put it on a place with a high accuracy. The last subtask is typical of the "peg-in-hole" operation.

Such a successive execution of fast transportation motion and precise guidance operation cannot be effectively solved, as a rule, by means of one control system and requires a combination of a rate control system (to carry out the first part of the task) and a position control system (to carry out the second, fine motion). This distribution of control method application follows both from theoretical considerations and experimental experience (Kan and Austin, 1990).

Just such situations demand combined position-rate systems of the tactical level, which work by alternately changing between the position and the rate modes (do not confuse with PD control systems of executing level that are used in robot servosystems as well as PI, PID and other types of control laws, where different control channels act simultaneously. In position-rate systems of tactical level control channels act sequentially). Switching from mode to mode is realized either by the HO or automatically. A possible general structure of the corresponding system is shown in Figure 4.42. The

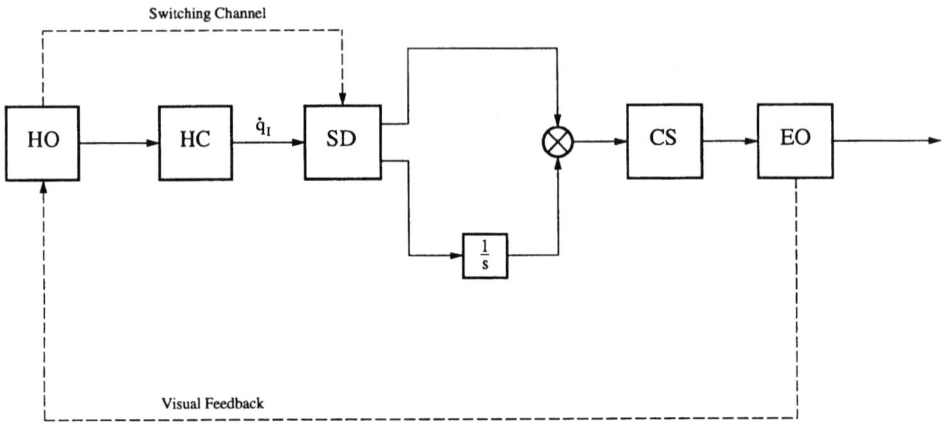

Figure 4.42. General block diagram of the tactical position-rate system. Here, SD is a switching device, and CS is a control system.

HO switches a SD according to the demands of a task performed (dashed line) either to a state of position control or to a state of rate control. Signals from these channels are summed, and after transformations in a CS enter inputs of the EO drives. It is important that the inverse kinematics are solved here for both control modes in the same computing device and using the same described kinematic algorithms.

When automatic switching of controlled channels takes place, a certain decisive rule is called into play, which is usually implemented by a non-linear device that just switches control channels. However, implementation of this device for these control methods is connected with additional conditions that lead to a complication of the entire control system.

The decisive rule is usually based on an analysis of rate characteristics of the HO control. In fact, we know that precise tasks are carried out with lower velocities and travel tasks are carried out with high velocities of the EO. Therefore, it would appear natural if there were relationships between high velocities of the HO arm motion and the rate control mode, and between low velocities and the position control mode. It is worth introducing a certain limitation on ID signal velocities $|\dot{\vec{x}}| \leq \dot{\vec{x}}_{max}$ that is used to control the switch. Thus, a possible implementation of the non-linear device (Slutski and Safontsev, 1983) can look as shown in Figure 4.43. Input signal x_i of each controlled channel goes at the beginning through a differentiator (1) and block (2) of the modulus determination. Components of setup vector $\dot{\vec{x}}_{max}$ from a source (3) of reference signals are compared with the modulus of velocities from the input of the block (2).

If a rate of change of any of these components is less than a correspond-

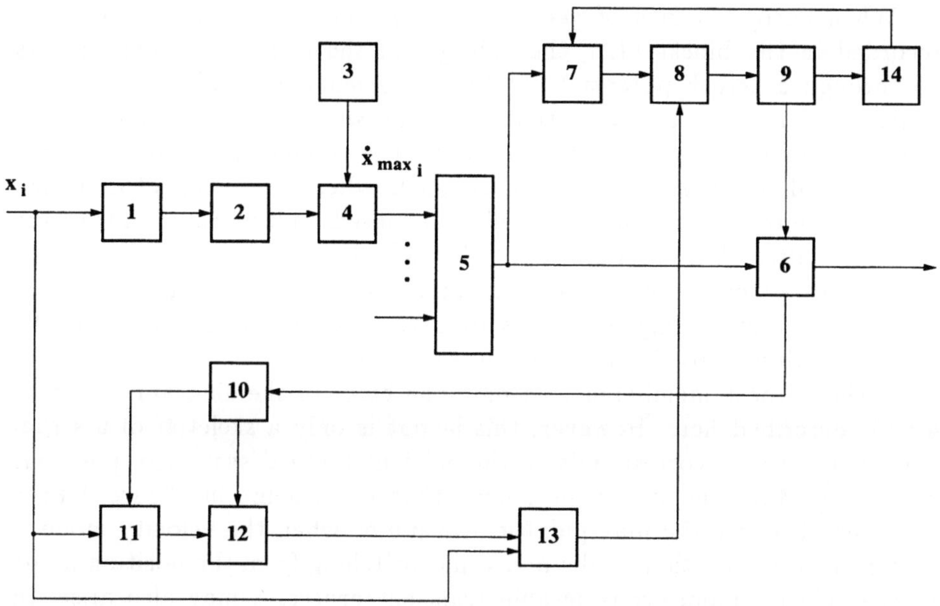

Figure 4.43. Block diagram of the switching device. Designations: 1 is a differentiator, 2 is a block of the modulus determination, 3 is a source of reference signals, 4 is a discriminator, 5 is an element "OR," 6 and 7 are triggers, 8 is a null-detector, 9 and 11 are switches, 10 and 14 are multivibrators, 12 is a memory block, and 13 is a summator.

ing threshold, no impulse goes from the discriminator (4) outputs to an input of an element (5) "OR". Triggers (6) and (7) remain in a state of waiting for the impulses from the discriminator (4) (they are automatically settled in this state when the control system is on). In this case, the triggers (7) transfer their switches (9) in open state through null-detectors (8). A control potential of trigger (6) carries out a switch of the system structure (Fig. 4.43). Thus, the device works at the position control mode.

When even one component of vector $\vec{\dot{x}}$ is in excess of the velocity setup, the discriminator (4) forms impulses that enter the trigger (6, 7) inputs through the block (5) "OR." These impulses transfer the triggers (6, 7) into their second stable state.

The voltage difference goes across the trigger (6) to the biased multivibrator (10). The latter forms a resolution pulse of short duration for the time the switches (11) are closed. The previous values of signals in memory blocks (12) are then canceled, and their new values which come from the system input, are recorded. When the resolution pulse from the outputs of the memory blocks (12) is over, new values of components of a recorded signal enter the inverting inputs of the summators (13), where they are summed with current values of the signal components.

When current values of vector $\vec{x}_I$ components are returned to those recorded in the blocks (12), the voltages of the summator (13) outputs go through a zeroth potential. At these moments, the null-detectors (8) generate short-duration pulses that enter the switch (9) inputs. Then, from the group of these switches controlled by the trigger (7), those turn out to be closed which are situated in channels where the setup of a change rate of a control signal was exceeded. The biased multivibrator (14) serves to delay pulses that enter its input through closed switches (9). It forms an impulse which, by its negative-going slope, returns the triggers (6, 7) to their state of waiting for pulses from the discriminator (4). Thus, the device is switched to the position control mode.

It should be remembered that only one block of the diagram in Figure 4.42 is described here. However, this layout is only a skeleton of a structural scheme that corresponds to the efficient tactical system of position-rate type. Moreover, it can be shown that the scheme in Figure 4.42 is inefficient because the manipulator arm jumps when the velocity mode is switched to the position mode; and while switching from the position mode to the velocity mode, considerable transient processes may also arise. In designing efficient control system, therefore, it is necessary to account for these features and to use special means to compensate for them. One possible version of the system that can then be obtained is shown in Figure 4.44.

In this version, the HO forms the command signals $u_1, ..., u_n$ with the help of the HC (Medvedev *et al.*, 1978). These signals enter the NCB and also the TLDB. Output signals of the nonlinear converters $\tilde{u_j}$ are computer-processed directly into the CD according to

$$s_j = A^{-1}\tilde{u_j},$$

where j is the DOF index, s_j is the input signal of computing device, and A^{-1} is the transformation inverse matrix.

From the CD, signals $s_1, ..., s_n$ enter an input of the CIB. This block contains the keys (K2,...,K4), which insert a necessary channel according to the commands of the TLDB. As this takes place, the signals s_i enter the inputs of the manipulator servodrives through an integrator or aperiodic unit. In the first case, the fast rate control occurs, and in the second case, the position control with high accuracy occurs.

The separate system blocks work as follows. The NCB ensures "the zeroth" values of signals u_j at the instant of switching from position mode to velocity mode of operation. Keys K1 are closed in the former and opened in the latter. If the command signal values do not exceed the threshold value, keys K1 and K2 are then closed, but keys K3 and K4 are opened.

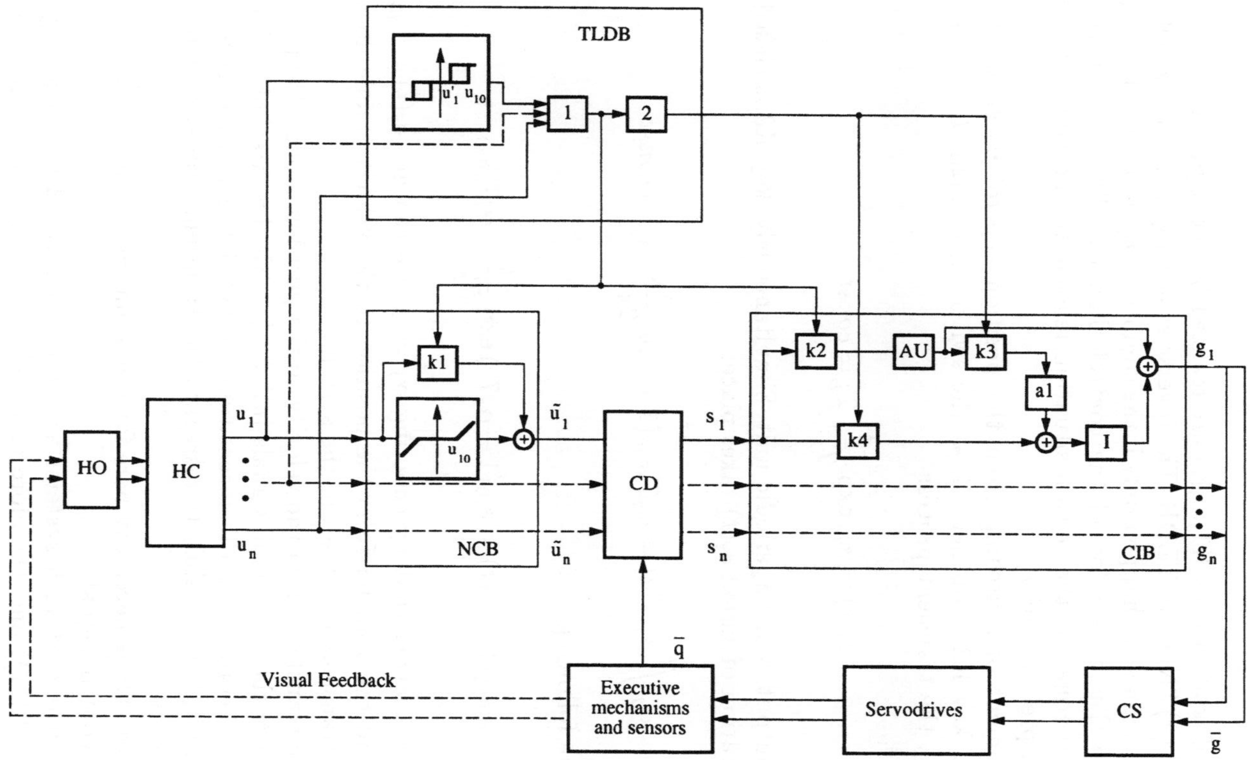

Figure 4.44. Detailed block diagram of the semiautomatic position-rate control system. Here NCB is a nonlinear converter block, TLDB is a block of threshold and logical devices, CD is a computing device, CIB is a communication and integration block and CS is a control system. Adopted from (Medvedev *et al.*, 1978); reproduced here by kind permission of Nauka, Moscow.

These key positions correspond to the position control. If only one value u_j stands out above the u_{j0} value, keys K3 and K4 are then closed, but keys K1 and K2 are opened. The coming-out scheme corresponds to velocity control.

The state of control keys K1,...,K4 are realized by the logical elements, (1) "OR," and (2) "NO," of TLDB. The inverse switching from the velocity mode to position one will happen only if the condition $|u_j| \leq u'_j, i = 1, ..., n$, takes place simultaneously for all the control channels. That is why the position control zone has the central situation relative to the manipulator gripper with a precision of u'_j.

At the moment of switching from the position control mode to the velocity mode, key K2 is opened. If, at the switching moment $t = 0$, the signal $s_j = s_{j0}$, the transient process

$$s_j(t) = s_{j0} \exp \frac{-t}{T_j}, \ T_j = const$$

arises at the output of the aperiodic unit. Simultaneously, key K3 is closed and the integration of signal $s_j(t)$ takes place:

$$\tilde{s_j}(t) = \alpha_j \int_0^t s_j(t) dt = \alpha_j T_j s_{j0}(1 - \exp \frac{-t}{T_j}), \ \alpha_j = const.$$

Signal g_j is the following:

$$g_j(t) = s_j(t) + \tilde{s_j}(t) = s_{j0}(1 - \alpha_j T_j) \exp \frac{-t}{T_j} + \alpha_j T_j s_{j0}.$$

From this, it is inferred that by choosing $\alpha_j = \frac{1}{T_j}$ we can get $g_j(t) = s_{j0}$ when $t > 0$, and the transfer from the position to the velocity mode does not cause the transient processes in the system.

The inverse transfer - from the velocity to the position mode - happens in the presence of the very small signals $|u_j| < u'_j$, with the threshold value u'_j determined according to the insensitivity zone of the control device. Thus, in this case, it is possible to disregard the transient processes which arise in the system.

Note that the CS is necessary here for the ordinary kinematic transformation of the control signals.

The problem is that such systems are very complicated (see also, e.g., (Melton *et al.*, 1971)) and therefore entail high costs and low reliability. This is the reason for the limited practical application of these systems in industrial robotics, though they have been well-elucidated in the scientific and patent literature.

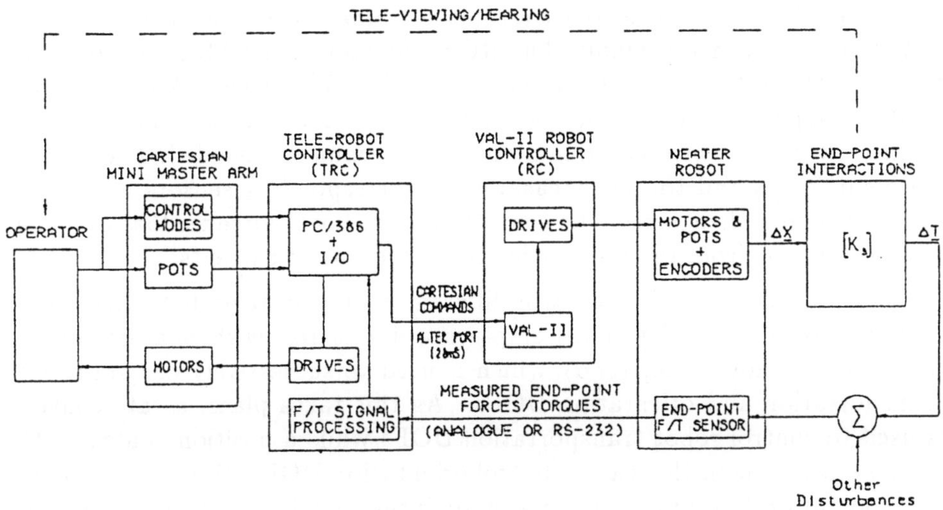

Figure 4.45. Block diagram of the remote manipulation system. Designations: RC is a robot controller, and TRC is a telerobot controller. From (Siva *et al.*, 1991).

Force control may also be combined with other control methods. It is used when the HO has to apply forces and moments to an object with constraints. Reactions of constraints occur then, and they may be either quite determined or unknown external forces. Semiautomatic force control systems can be created with all their above-mentioned advantages (Bejczy, 1994; Medvedev *et al.*, 1978), and they can be combined with remote methods of position or rate control (Kuleshov *et al.*, 1986).

4.5.2. SEMIAUTOMATIC CONTROL IMPLEMENTATION

Depending on an ID type, there are two possible ways to implement SAC systems: when each axis of the HC constantly controls one robot axis, and when one HC axis controls, in turn, different robot axes. The first way is usually in the use when a multi-axis HC is applied. It is also in the use when, for instance, two 3-axis HCs are used in parallel to control a 6-DOF manipulator. The second way is used, as a rule, when a HC with small number of axes is applied to control different groups of robot DOFs.

In the first case, when only one input sensor of a HC is used to set one control signal for a certain robot drive, the control system has a traditional structure. Figure 4.45 shows a general block diagram of a control system using CmMA HC (Fig. 4.7) as ID (Siva *et al.*, 1991). The system was intended to control the NEATER (Nuclear Engineered Advanced Telerobot) robot, which was fitted with a 6-axis force/torque sensor to measure endpoint forces and torques in the tool coordinate frame. Potentiometers ("pots") initiated input control signals. The RC was a standard VAL-II control system, and the TRC was a PC-based system. The force feedback control was made to bypass the VAL-II RC so that communication speed limits, which greatly affect the control system performance, could be avoided. The slave's endpoint forces and torques measured by corresponding sensors were conditioned before their use in the control system. Local force feedback on the ID was derived from the force reflecting servo motor currents.

When an ID is used to control different groups of telerobot DOFs, the system shown here (Fig. 4.46) employs a 4-axis HC with bilateral capabilities (Kuleshov *et al.*, 1986), which is used to successively control robot transportation and orientation DOFs. As this takes place, a rate control is used to control robot transportation DOFs while a position control with force reflection is in the use to control orientation DOFs. While performing transportation operations, the robot orientation DOFs are immobile and the HO sets the direction of a spatial vector of the robot endpoint velocity with the help of the revolute kinematic pairs II, III, and IV, which determine a vector of angle coordinates $\vec{B} = [\beta_2\beta_3\beta_4]$. The HO then uses a linear displacement of the kinematic pair I from the zero position to set a modulus and a direction of this vector. A computer forms signals for robot transportation DOFs by solving inverse kinematics based on the use of the pseudoinverse of the Jacobian matrix A:

$$[\dot{q}_1 \ldots \dot{q}_m]^T = A^{-1}(\vec{q})\beta_1\vec{B}.$$

While performing orientation operations, the robot transportation DOFs are braked by either program or mechanical means. ID kinematic pairs II, III, and IV then form bilateral systems with robot corresponding orientation DOFs and pair I forms the bilateral system with the EO drive that develops the assembly force.

One feature of described control systems has to be mentioned: the requirement for system signal scaling, which is necessary for both control and sensing signals. Thus, it has been established that, for microsurgerical telerobots, for example, the following requirements have to be satisfied (Salcudean and Yan, 1994):
- downward displacement scaling from master to slave in order to increase the resolution of the surgerical tool and attenuate physiological tremor;
- upward force scaling from slave to master to increase a surgeon's sensitivity.

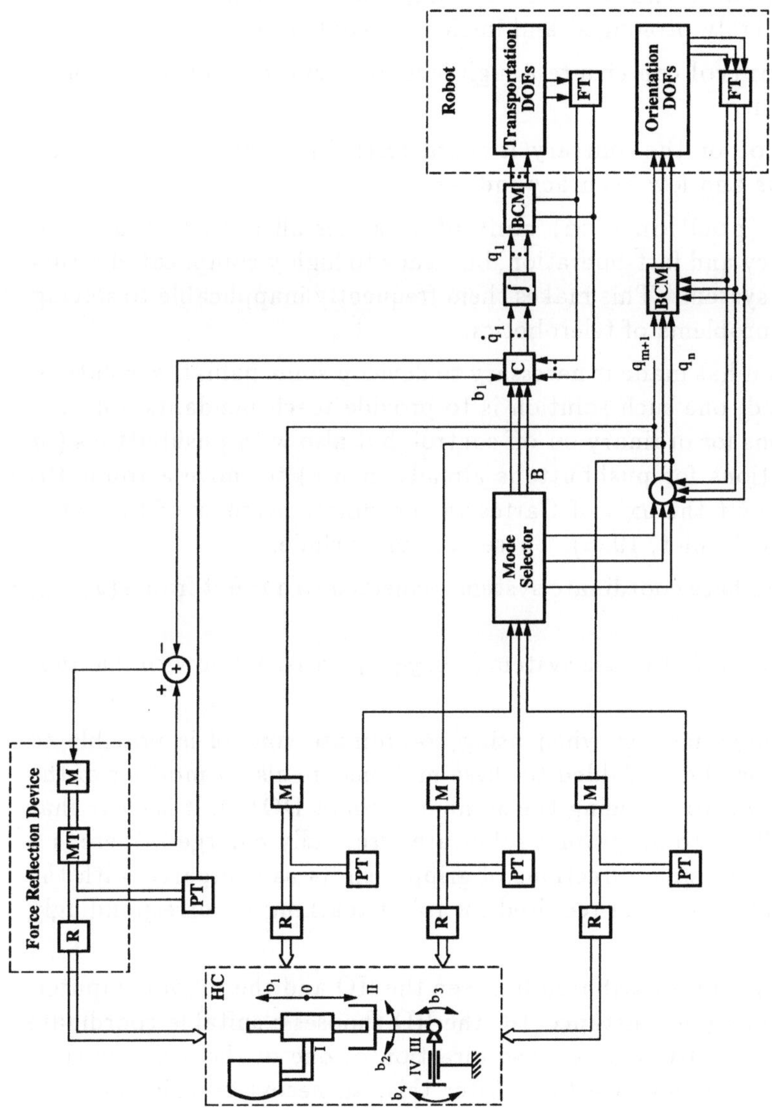

Figure 4.46. Block diagram of a remote manipulation system with successive control of different groups of robot DOFs from one HC. Designations: C is a computer; R is a reducer; M, a drive motor; MT, a moment transducer; PT, a position transducer; FT, a feedback transducer; and, BCM, a block of the drive motor control. Adopted from (Kuleshov *et al.*, 1986); reproduced here by kind permission of Mashinostroenie, Moscow.

4.6. Co-ordinate Control

The methods of remote control incorporate the following disadvantages, which have only been partly noted above:

On-off control is characterized by simplicity and sufficient accuracy, but leads to a highly prolonged and ineffective control process.

Position control can ensure a high level of accuracy, but a low speed of task fulfillment.

Rate control, on the contrary, is characterized by a high level of speed of teleoperations and low accuracy indices.

Combined (position-rate) control possesses all the advantages relating to accuracy and fast-operation, but leads to highly complicated structures of control systems. This makes them frequently inapplicable to solving some practical problems of telerobotics.

This situation has made it necessary to develop some palliative solution. As already noted, one such solution is to provide teach pendants not only with pushbuttons for ordinary on-off control, but also with pushbuttons (or additional functions for pushbuttons already in use) to move a robot EE along to and about the axis of Cartesian coordinate systems of the robot arm (Vertut and Coiffet, 1985). These systems include:

- a motionless base coordinate system connected to a robot frame (x_0, y_0, z_0) (Fig. 4.47);

- a movable tool coordinate system (x_G, y_G, z_G) associated with manipulator EE.

Thus, a gripper motion when using co-ordinate control is possible to six directions (Fig. 4.47). Added to these motions are also n motions of the ordinary on-off control (n being the number of robot DOFs). It is clear that the HO possibilities to program a robot are essentially enlarged, there is a greater choice of possible directions of gripper shifts as compared with the on-off control, and the time required for robot teaching is correspondingly reduced.

Calculation load is distributed between the HO and the robot computer. While programming, for instance, IR, the HO chooses a suitable coordinate system; that is, the HO chooses the direction of one of the axes which is the closest to the target direction. After this, all necessary calculation to transform coordinates is carried out by the robot computer.

It may be seen from our presentation that the method of co-ordinate control is somewhat intermediate mean between on-off and semiautomatic controls. It differs from the latter by greater simplicity but also smaller flexibility and effectiveness.

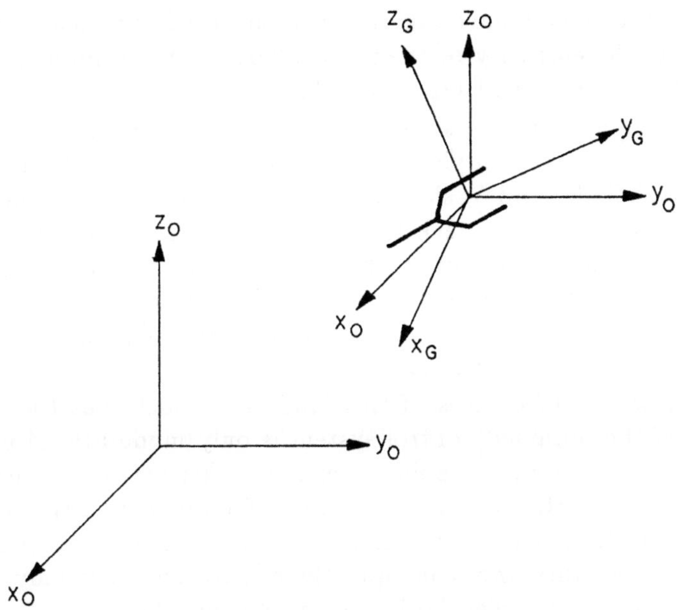

Figure 4.47. Coordinate systems used for co-ordinate control.

4.7. Discussion

It is our belief that the on-off control has earned its wide application essentially, because of the disadvantages of existing remote control methods. This on-off method may be sometimes useful for controlling robots; however, only in combination with other more effective control methods. This resembles the situation where we can always carry out simple arithmetic calculations on the computer display for ourselves, but computer software has to be invoked to solve more complicated cases.

Among the more promising control approaches, is, first, the MSSs. As noted, this method also has some above negative aspects, but it is used in practice because of its relatively simple implementation.

The method using the FM leads to the construction of systems that have some advantages over conventional MSSs. The developed master is very simple, light, reliable, and cheap. This is because kinematic pairs and moving links are absent. Consequently, the need to balance master masses is eliminated, leading to lighter construction and an overall decrease in dimensions and cost. Such devices are also universal, because, by placing additional gauges on the spring, it is possible to use them to control a broad spectrum of different kinematic structures of manipulators.

Another point - and this is the main aspect of the device, in our opinion

- is that it can be successfully used to control redundant manipulators. The FM permits a master-slave regime to be achieved for such manipulators. In this case, the elasticity of the master flexible spring imposes additional constraints on the extra DOF of the slave.

Currently, many masters are provided by force reflecting, which significantly decreases task completion time (Jansen *et al.*, 1992; Matsuhira *et al.*, 1993; Yokokohji and Yoshikawa, 1994). We have hardly touched on the problem of force feeding back to the FM, but this problem may also be solved in context of this approach. Thus, passive and active reflection of forces from the slave may be organized while developing these techniques. We have shown the possibility of this approach by designing a corresponding variety of the master (Figs. 4.34 and 4.35).

One drawback of devices of this kind is the fact that the positioning accuracy of the obtained control system is only moderate. Moreover, this accuracy varies for different points over the workspace - a common property of all MSSs. The HO usually compensates for this inaccuracy of MSSs by using a visual feedback. At the same time, our static evaluations indicate that mnemonicability and, consequently, control accuracy, only deteriorate moderately, while the telerobotic system remains efficient.

As for the SAC systems, they may be characterized as the most promising system for lightening the HO's workload. They permit the reduction of physical and mental loads on the HO, their IDs have very little weight and overall dimensions, and they require only small shifts of the HO's arm.

A great problem of industrial robotics is that methods of SAC are very seldom used to control the IRs while teaching them. Individual attempts to solve this problem, by using simpler means based on position semiautomatic control ideas (Brantmark *et al.*, 1982) have not done the job, and they are not widely realized under industrial conditions. One reason was put forth by Parsons (1992): customers complained that joysticks were damaged when a teach pendant dropped; but this circumstance is not a reason, but an event. The reason is perhaps that existing remote systems do not ensure required levels of precision, fast-operation, and reliability. Because the figure of 25% of teaching time reduction obtained by ASEA is small enough to be ignored in most situation, there is less incentive to take care not to drop the teach pendants; however, if the figure were greater (e.g., up to 100%), the incentive to take better care of the equipment might be correspondingly greater. In fact, there is a great deal of valuable equipment (e.g., metrological instruments) that should never drop and do not. Therefore, our conclusion is that present-day teaching equipment is not treated as valuable from the standpoint of the users. Other reasons, therefore, become decisive when choosing a method of robot teaching.

An important practical problem of SAC application concerns choosing

the type and number of HCs to control a multi-DOF robot. O'Hara and Olsen (1988) have performed investigations without finding any advantage of two three-axis HCs over the six-axis controller. They concluded that the integration of all axes into a single controller is desirable because of the following reasons:
- equipment reduction (a two-controller workstation provides the most redundancy);
- utilization of limited vehicle space (one controller uses less space);
- free-hand operations (the six-axis controller frees a HO's other hand to operate secondary functions).

Among other problems that have not been properly studied is the dynamic coupling problem between the HO and manipulator in MSS and, even less, in SAC systems. This subject has not been adequately explored and only now is beginning to be sufficiently developed (Yokokohji and Yoshikawa, 1994).

The outlined demerits of existing remote control methods lead us to conclude that further development and perfection of on-line control is necessary due to the indicated advantages; e.g., simplicity of realization, lower cost, versatility, and no need for high qualification of personnel (Hartley, 1985; Semrad, 1982; Weck and Lauffs, 1977).

References

Andary, J.F. and Spidaliere, P.D. (1993) The development test flight of the flight telerobotic servicer: Design description and lessons learned, *IEEE Trans. on Robotics and Automation*, **Vol. 9 no. 5**, pp. 664-674.

Asano, K. (1984) Control system for a multi-joint inspection robot, *Proc., Rob. and Remote Hand. Hostile Environ.*: Nat. Top. Meet., Gatlinburg, TN, La Grange Park, Ill, pp. 375-382.

Bejczy, A.K. (1994) Toward advanced teleoperation in space, in: S.B. Skaar and C.F. Ruoff (eds.), *Teleoperation and Robotics in Space.* American Institute of Aerodynamics and Astrodynamics, Progress in Aeronautics and Astronautics Series, AIAA, Washington D.C., pp. 107-138.

Brantmark, D., Lindquist, A., and Norefors, U.G. (1982) Man-machine communication in ASEA's new robot controller, *ASEA J.*, **Vol. 55 no. 6**, pp.145-150.

Burdea, G. (1989) Human/Machine Interaction in Telerobotic Dextrous Feedback, *Symp. on Dynamics and Control of Biomechanical Systems*, 1989 ASME Winter Annual Meeting, pp. 65-69.

Burdea, G. and Coiffet, P. (1994) *Virtual Reality Technology.* John Wiley & Sons, New York.

Collins, C.L. and Long, G.L. (1995) The singularity analysis of an in-parallel hand controller for force-reflected teleoperation, *IEEE Trans. on Robotics and Automation*, **Vol. 11 no. 5**, pp. 661-669.

Dietrich, J. and Plank, G. (1988) Optoelectronic system housed in a plastic sphere, US patent no. 4,785,180.

Djamalov, N.K. and Slutski, L.I. (1986) Mnemonicability level evaluation for master-slave manipulator with elastic setting organ, in *Problems of Mechanism and Machine Theory*: Collection of scientific papers. Kazakh State University, Alma-Ata, pp. 144-

148 (in Russian).

Djoldasbekov, U.A., Konstantinov, M.S., Markov, M.D., and Slutski, L.I. (1983) Executing mechanism of a robot-manipulator, USSR author's certificate no. 1081919 (in Russian).

Djoldasbekov, U.A., Slutski, L.I., and Djamalov, N.K. (1990) Master manipulator, USSR author's certificate no.1563962, *Bulletin of Inventions* , **no.18** (in Russian).

Efremov, E.V. (1978) Effectiveness increase of an operator-manipulator system by means of feedback vibrotactile channel, *Institute for the Study of Machines of the USSR Academy of Sciences. Ph. D. Dissertation.* Moscow (in Russian).

Farry, K.A. and Walker, I.D. (1993) Myoelectric teleoperation of a complex robotic hand, *Proc. of IEEE Intern. Conf. on Robotics and Automation*, **Vol.3**, pp. 502-509.

Fischer, P., Daniel, R. and Siva, K.V. (1990) Specification and design of input devices for teleoperation, *Proc. of IEEE Intern. Conf. on Robotics and Automation*: Cincinnati, Ohio. IEEE Computer Society Press, Los Alamitos, California, pp. 540-545.

Gladovski, T., Pedram, H. and Shamash, Y. (1992) Human factors in robot teach programming, in M. Rahimi and W. Karwowski (eds.), *Human-Robot Interaction.* Taylor & Francis, London, pp. 16-46.

Handlykken, M. and Turner, T. (1980) Control systems analysis and synthesis for a six degree-of-freedom universal force-reflecting hand controller, in *Proc. of 19th IEEE Conf. on Decision and Control*, Albuquerque, NM, Dec. 10-12, pp. 1197-1205.

Hannaford, B., Wood, L., McAffee, D.A., and Zak, H. (1991) Performance evaluation of a six-axis generalized force-reflecting teleoperator, *IEEE Trans. on Systems, Man, and Cybernetics* , **Vol.21 no.3**, pp. 620-633.

Hartley, J. (1985) Robotique: vive l'entrainement direct!, *Usine nouv.*, **43**, p. 87 (in French).

Heindl, J. and Hirzinger, G. (1986) Device for programming movements of a robot, US patent no. 4,589,810.

Hirzinger, G. (1993) Multisensory shared autonomy and telesensor programming - Key issues in space robotics, *Robotics and Autonomous Systems*, **11**, pp. 141-162.

Hollerbach, J.M. and Lokhorst, D.M. (1995) Closed-loop kinematic calibration of the RSI 6-DOF hand controller, *IEEE Trans. on Robotics and Automation*, **Vol. 11 no. 3**, pp. 352-359.

Hunt, K.H. (1983) Structural kinematics of in-parallel-actuated robot-arm, *ASME J. of Mechanisms, Transmissions, and Automation in Design*, **Vol. 105**, pp. 705-712.

Iwata, H. (1993) Pen-Based Haptic Virtual Environment, *Proc. of IEEE Virtual Reality Annual Intern. Symp.*, Seattle, WA, September, pp. 287-292.

Jacobus, H.M., Riggs, A.J., Jacobus, C.J., and Weinstein, Y., (1992) Implementation issues for telerobotic handcontrollers: human-robot ergonomics, in M. Rahimi and W. Karwowski (eds.), *Human-Robot Interaction.* Taylor & Francis, London, pp. 285-314.

Jansen, J.F., Kress, R.L., and Babcock, S.M. (1992) Controller design for a force-reflecting teleoperator system with kinematically dissimilar master and slave, *ASME J. of Dynamic Systems, Measurement, and Control*, **Vol. 114**, pp. 641-649.

Kan, E.P. and Austin, E. (1990) The JPL telerobot teleoperation system, *Intern. J. of Robotics and Automation*, **Vol. 5 no. 1**, pp. 27-31.

Klafter, R.D., Chmielewski, T.A., and Negin, M. (1989) *Robotic Engineering: an Integrated Approach.* Prentice Hall, Englewood Cliffs, NJ.

Kobrinskii, A.E., Bolkhovitin, S.V., Voskoboinikova, L.M., Ioffe, D.M., Polyan, E.P., Popov, B.P., Slavutski, Ya.L., Sysin, A.Ya., and Yakobson, Ya.S. (1960) Problems of bioelectric control, *Proc. of 1st Intern. Congress of IFAC*, Moscow.

Kobrinskii, A.E., Statsenko, V.V., and Tyves, L.I. (1975) Mnemonicability level evaluation of the master-slave manipulator with disconcordance of the input and executing organs, *Mashinovedenie*, **no. 2**, pp. 3-7 (in Russian).

Kochan, A. (1996) Hanover - Robots under control, *Industrial Robot*, **Vol. 23 no. 4**, pp. 31-33.

Kuleshov, V.S., Lakota, N.A., Andrunin, V.V., Belousov, B.N., Gorbatsevich, E.D., Dorokhov, V.P., Egorov, I.N., Moiceenkov, B.A., Poduraev, Yu.V., Shvedov, V.N., and Yushchenko, A.S. (1986) *Remotely Controlled Robots and Manipulators*. Mashinostroenie, Moscow (in Russian).

Lawrence, D.A. (1992) Designing teleoperator architectures for transparency, *Proc. of IEEE Intern. Conf. on Robotics and Automation*: Nice, France, pp. 1406-1411.

Long, G.L. and Hao, F. (1995) Singularity resolution and force-reflection for an optimal geometry in-parallel planar manipulator, *Proc. Ninth World Congress on the Theory of Machines and Mechanisms*, **Vol.3**, Milan, Italy, pp. 1734-1739.

Matsuhira, N., Asakura, M., Bamba, H., and Uenohara, M. (1993) Development of an advanced master-slave manipulator using a pantograph master arm and a redundant slave arm, *Proc. of IEEE RSJ Intern. Conf. Intell. Rob. Syst.*: Yokohama. IEEE Service Center, Piscataway, NJ, pp. 1653-1658.

Medvedev, V.S., Leskov, A.G., and Yushchenko, A.S. (1978) *Control Systems of Manipulation Robots*. Nauka, Moscow (in Russian).

Melton, F., Rose, G.R., and Hedin, R.S. (1971) Position control manipulator, USA patent no. 3280991.

Nechyba, M.C. and Xu, Y. (1995) Human-robot cooperation in space: SM^2 for new space station structure, *IEEE Robotics and Automation Magazine*, **Dec.**, pp. 4-11.

Nielsen, J. and Roth, B. (1996) The direct kinematics of the general 6-5 Stewart-Gough mechanism, in J. Lenarcic and V. Parenti-Castelli (eds.), *Recent Advances in Robot Kinematics*. Kluwer Academic Publishers, Dordrecht, pp. 7-16.

O'Hara, J.M. and Olsen, R.E. (1988) Control device effects on telerobotic manipulator operations, *Robotics and Autonomous Systems*, **Vol. 4 no. 1**, pp. 5-18.

Parsons, H.M. (1992) Remote-control units for industrial robots, in M. Rahimi and W. Karwowski (eds.), *Human-Robot Interaction*. Taylor & Francis, London, pp. 284-314.

Ponomariov, S.D. and Andreeva, L.E., (1980) *Calculations of Elastic Elements of Machines and Instruments*. Mashinostroenie, Moscow (in Russian).

Raghavan, M. and Roth, B. (1990a) Inverse kinematics of the general 6R manipulator and related linkages, *Mechanism Synthesis and Analysis*, **DE-Vol. 25**. ASME, pp. 59-65.

Raghavan, M. and Roth, B. (1990b) Kinematic analysis of the 6R manipulator of general geometry, in H. Miura and S. Arimoto (eds.), *Robotics Research, Proc. of Fifth Int. Symp.*. MIT Press, pp. 263-270.

Raghavan, M. and Roth, B. (1995) Solving polynomial systems for the kinematic analysis and synthesis of mechanisms and robot manipulators, *Special 50th Anniversary Design Issue, J. of Mechanical Design and J. of Vibration and Acoustics, Trans. ASME*, **Vol. 117(B)**, pp. 71-79.

Reboulet, C., Plihon, Y., and Briere, Y. (1995) Interest of the dual hybrid control scheme for teleoperation with time delays, *Preprints of the Fourth Intern. Symp. on Experimental Robotics, ISER'95*: Stanford, California, USA, pp. 1-6.

Saenger, E.L. and Pegden, C.D. (1973) Terminal pointer hand controller and other recent teleoperator controller concepts: technology summary and application to Earth orbital missions, in E. Heer (ed.), *Remotely Manned Systems*. California Institute of Technology, pp.327-336.

Salcudean, S.E. and Yan, J. (1994) Towards a force-reflecting motion-scaling system for microsurgery, *Proc. of IEEE Intern. Conf. on Robotics and Automation*: San-Diego, USA, pp. 2296-2301.

Semrad, H. (1982) Vorbetrachtungen zur Programmierung von industrierobotern aus der sicht der anwenders, *Fertigunstechnik und Betrieb.*, **Vol.32 no. 11**, pp.670-674, 642 (in German).

Sheridan, T.B. (1992) *Telerobotics, Automation, and Human Supervisory Control*. The MIT Press, Cambridge, MA.

Siva, K.V., Fischer, P.J., Brown, M.H., and Abel, E. (1991) The development of a bilateral input device for use in teleoperation, in M. Jamshidi and P.J. Eicker (eds.), *Robotics and Remote Systems. Proc. Fourth ANS Topical Meeting on Robotics and Remote Systems*, pp.621-632.

Slutski, L.I. (1986) Master manipulator for a slave arm, USSR author's certificate no.1229028, *Bulletin of Inventions*, **no. 17** (in Russian).

Slutski, L.I. (1995) Universal master arm for teleoperation, *Proc. 7th Intern. Conf. on Advanced Robotics*, **Vol. II**, Spain, pp. 577-583.

Slutski, L.I. and Safontsev, E.A. (1983) Device for manipulator position-rate control, USSR author's certificate no.991366 *Bulletin of Inventions*, **no. 3** (in Russian).

Thring, M.W. (1983) *Robots and Telechirs: Manipulators with Memory; Remote Manipulators; Machine Limbs for the Handicapped.* Ellis Horwood Ltd., Chichester.

Todd, D.J. (1986) *Fundamentals of Robot Technology.* Kogan Page, London.

Tomovic, R. (1960) Human hand as a feedback system, *Proc. 1st Intern. Congress of IFAC*, Moscow.

Tomovic, R. (1969) On man-machine control, *Automatica*, **no. 5**, pp. 401-404.

Vertut, J. and Coiffet, P. (1984) *Teleoperation and Robotics: Evolution and Development*, Vol. 3A. Prentice-Hall, Englewood Cliffs, NJ.

Vertut, J. and Coiffet, P. (1985) *Teleoperation and Robotics: Applications and Technology*, Vol. 3B. Prentice-Hall, Englewood Cliffs, NJ.

Weck, M. and Lauffs, H.G. (1987) Sensoren beschleunigen On-Line-Roboterprogrammierung, *Ind.- Anz.*, **Vol.109 no. 8**, pp. 19-22 (in German).

Wilt, D.R., Pieper, D.L., Frank, A.S., and Glenn, G.G. (1977) An evaluation of control modes in high gain manipulator systems, *Mechanism and Machine Theory*, **Vol.12**, pp. 373-386.

Wurll, P. und Fritz, H. (1985) Steuerstab, Deutsches patent no. DE 3404047 A1 (in German).

Yokokohji, Y. and Yoshikawa, T. (1994) Bilateral control of master-slave manipulators for ideal kinesthetic coupling - formulation and experiment, *IEEE Trans. on Robotics and Automation*, **Vol. 10 no. 5**, pp. 605-620.

CHAPTER 5

CONTROL WITH PARAMETER ADJUSTMENT IN THE REMOTE MANIPULATORS

Basis of the Organization of a New Type of Remote Systems

5.1. The Experimental Basis of an Adjustment Method

The application of the criteria for accuracy and fast-operation of a manipulation performance during system quality analysis (Subsection 3.2.3) has indicated their utility in evaluating the abilities of various operators controlling the same manipulator, as well as the quality of different manipulators controlled by the same operator.

The use of these criteria for designing remote control for present-day machines is based on the fact that alteration of parameter values of a controlled manipulator influences system performance quality. The determination of this influence was the purpose of the experimental investigation performed with the help of semi-natural simulation methods described in Section 3.4.

5.1.1. EXPERIMENT ORGANIZATION

The remote control process of the position-type two-coordinate manipulator has been studied. During the investigation of the layout shown in Figure 5.1, the HO controlled the plane Cartesian manipulator linear model (M), which was implemented on an analog computer, using a two-coordinate input device (SD) to set control signals.

For each control channel, a dynamic manipulator model was adopted for the oscillating unit and was performed with the same transfer function

$$W_x(p) = W_y(p) = \frac{k}{T^2p^2 + 2T\xi p + 1},$$

where T is the period of free vibrations of the model, ξ is the damping coefficient, and k is the model gain. The output point of the EO model was displayed with a mark on the oscilloscope (O) screen. The control task for the HO was fast and precise tracing of a prescribed closed curve (specifically, a circle) plotted on the oscilloscope screen using visual feedback. According to the remote system quality evaluation approach described in Chapter 3,

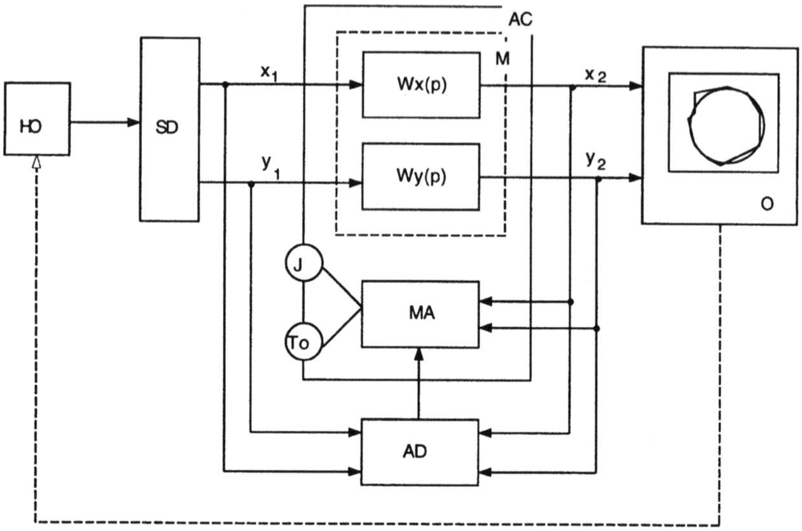

Figure 5.1. Layout of the experimental unit for remote system semi-natural simulation. Designations: M is a manipulator model, AC is an analog computer, SD is an input device, O is an oscilloscope, MA is a measuring apparatus, and AD is an automatic device. From (Slutski and Coiffet, 1996). ©Cambridge University Press.

the performed task is related to the type O_3 (an object transportation on a prescribed trajectory) and the criterion estimate is here C_5 (to minimize both criteria of accuracy and operation time).

During task execution, the measuring apparatus (MA) was switched on, with the help of the automatic device (AD), and recorded the total time T_0 of the control cycle and the system integral quality index J (Subsection 3.2.3):

$$J = \int_0^{T_0} [\rho(t) - \rho_r(t)]^2 dt \, , \tag{5.1}$$

where ρ is the modulus of the radius vector of the trajectory realized by the HO and ρ_r is the modulus of the reference trajectory radius vector. Index (5.1) is approximately equal (with an accuracy of co-multiplier (Slutski, 1972)) to the complex criterion

$$J \approx T_0 \cdot \Phi = T_0 \int_{\phi_0}^{\phi_{T_0}} [\rho(\phi) - \rho_r(\phi)]^2 d\phi \, , \tag{5.2}$$

where Φ is an estimate of the integral error of curve tracing, and ϕ is the polar angle.

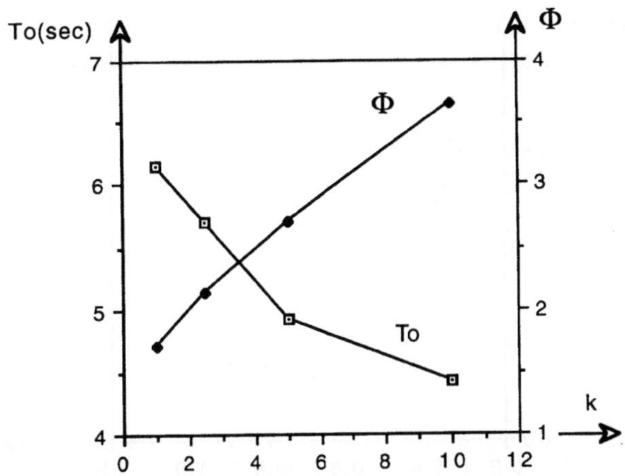

Figure 5.2. Dependences of task completion time T_0 and integral error Φ from manipulator gain k.

5.1.2. ANALYSIS OF EXPERIMENT RESULTS

Using recorded results T_0 and J, the calculation of accuracy index Φ was produced. The end goal of the investigation at this stage was to obtain the dependencies

$$J = f_1(k),\ J = f_2(T),\ J = f_3(\xi),$$
$$T_0 = f_4(k),\ T_0 = f_5(T),\ T_0 = f_6(\xi),$$
$$\Phi = f_7(k),\ \Phi = f_8(T),\ \Phi = f_9(\xi).$$

We present here these dependences for two main criteria, T_0 and Φ. First, the dependences of these quality criteria from manipulator gain are presented (Fig. 5.2). This indicates a rather determined connection between these values: the task completion time depends almost inversely on the robot gain, and the task error depends on this parameter in approximately direct proportion.

Other dynamic manipulator parameters ξ and T variously influence the remote control quality (Figs. 5.3 and 5.4).

Based on the above dependences, the relative sensitivity coefficients were calculated for each quality criterion J, Φ, and T_0, with respect to each of parameters k, T, and ξ. The corresponding sensitivity coefficient values, based on statistically processed experimental data, are presented in Table 5.1 (Slutski and Safontsev, 1983).

It is evident from the data in Table 5.1 that manipulator model gain k was the most influential of all the criteria. At the same time, the parameters ξ and T variously influenced the criteria J, Φ, and T_0. The principal influ-

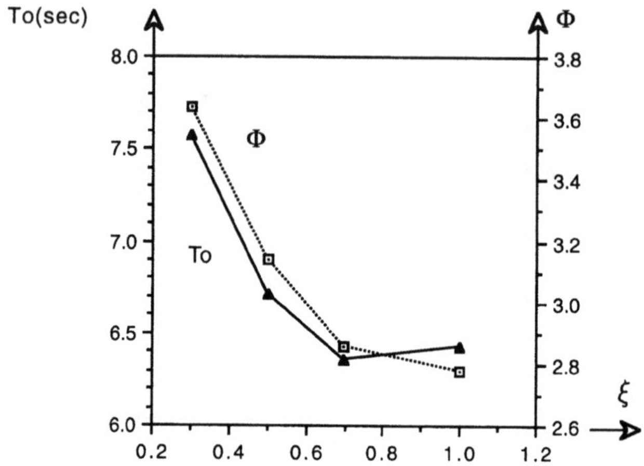

Figure 5.3. Dependences of task completion time T_0 and integral error Φ from manipulator damping coefficient ξ.

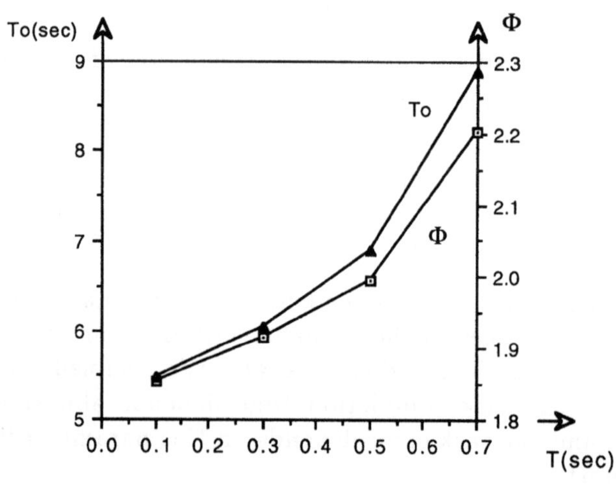

Figure 5.4. Dependences of task completion time T_0 and integral error Φ from manipulator period T of free vibrations.

ence on the task performance speed is vibration period T, and the principal influence on the system accuracy and complex criterion J is the damping coefficient ξ.

From the above data, it also transpires that control process quality is best when $\xi \to 1$ and $T \to 0$. This result coincides with the conclusion that the ideal remote system must be transparent(Handlykken and Turner,

TABLE 5.1. Values of the relative sensitivity coefficients

Quality Criterion	T_0	Φ	J
Varied Parameter			
k	2.49	2.18	8.75
ξ	0.9	0.7	5.99
T	2.4	0.07	0.9

1980). Thus, an equivalent passage from the pair of criteria $I_1 = T_0, I_2 = \Phi$ to another pair of dynamic criteria can be performed. These are already related to the controlled object parameters:

$$Q_1 = T \to min; Q_2 = |\xi - 1| \to min\,. \tag{5.3}$$

In our works, we used these dependences for the synthesis of a telerobot control system of executing level; i.e., robot servodrives (Slutski, 1995; Slutski *et al.*, 1983). These results are not presented here because they are not connected with the principal subject of the present book; i.e., control at telerobot tactical level.

However, it is important to note that, if there is a need to accomplish work in an optimal regime, the manipulator gain is the first parameter that should be tuned. One way to perform such tuning is by setting the optimal gain values for all the controlled manipulator channels. This may be done by adjusting the control system amplifiers beforehand.

For such adjustment of the gain value, one has to be sure that the optimal parameter value is approximately constant in the HO's work and that there is no prevailing tendency for it to change over time. To ascertain this, another experimental study was performed.

5.2. Optimal Value Tuning of Parameters

The work quality study of a telerobotic system with manipulator gain tuning was based on the experimental methods described in the previous section (Fig. 5.1). An inertialess model with rate control using equation

$$\dot{y}_2 = \eta \cdot y_1 \quad , \quad \dot{x}_2 = \eta \cdot x_1 \, ,$$

(where $\eta[\sec^{-1}]$ is the gain of the rate manipulator, x_1 and y_1 are input signals from an ID, x_2 and y_2 are output signals of the manipulator model) was studied in this case.

Under investigation were six values of parameter η: 0.135; 0.2; 0.675; 1.35; 2.025; and 3.375. The efficiency criterion (5.2) was calculated when the gain tuning was adjusted based on the mentioned segment where the optimal value of the parameter

$$\eta_0 = \arg J(\eta)|_{J=J_{min}}$$

was determined.

To determine the optimal gain value $\eta=\eta_0$ and its change in time, a permanent scan was performed on the enumerating values of parameter η within the selected segment. The scan was performed according to alternate direct and reverse directions in order to avoid large jumps in the gain that would have caused the operator's work quality to deteriorate due to his loss of control skill. For each gain value, a reference trajectory tracing was executed. For scanning each cycle, the optimal value η_0 was determined. Thereby, a set of characteristic results was obtained, despite all the random errors accompanying the HO work.

The alteration of the optimal gain η_0 vs. the number N of the tracing cycle (scan) is shown in Figure 5.5 (the continuous line). It can be seen from this diagram that the optimal value of $\eta_0 = 3.375\,\text{sec}^{-1}$ remained about constant, with some exceptions, during almost the entire experiment. It is necessary to consider the fact that the described investigation was accomplished over a lengthy period (up to six hours). Thus, it is possible to conclude that a certain optimal value of the manipulator gain for every operator remains fairly constant during the entire working day.

In the experimental process, criterion (5.2) with the simplified accuracy component,

$$J^* = T_0 \int_0^{T_0} [\rho(t) - \rho_r(t)]^2 dt \; , \tag{5.4}$$

was also calculated. Another sequence of the optimal values η_0 corresponds to this criterion (the dotted line in Fig. 5.5). It is important to note that the different criteria (5.2) and (5.4) have different optimal values of η_0.

The effectiveness of the chosen optimal manipulator parameter is also important. It is necessary to clarify whether this parameter will really affect a system quality increase. To determine this, the values of the complex criterion J (in relative units), corresponding to the optimal parameter value $\eta_0 = 3.375\,\text{sec}^{-1}$, were selected from the results of the tracing of each cycle. Their sequence vs. the tracing cycle number N is shown in Figure 5.6 (a continuous line).

The values of criterion J for the nearest approach to optimal gain η_0 value, $\eta_1 = 2.025\,\text{sec}^{-1}$ were also selected, as well as shown in Figure 5.6 (a dotted line). It is clear that line $J_{\eta_0}(N)$ has settled down almost everywhere

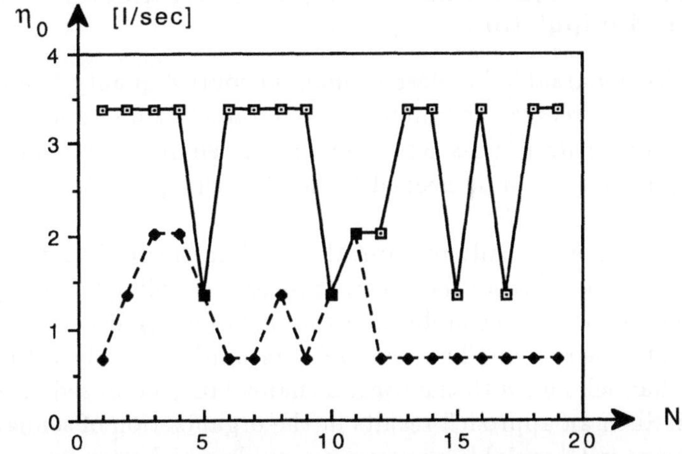

Figure 5.5. Alteration of optimal gain η_0 vs. trial number N for different optimality criteria: J (5.1) - a continuous line, J^* (5.4) - a dotted line. From (Slutski and Coiffet, 1996). ©Cambridge University Press.

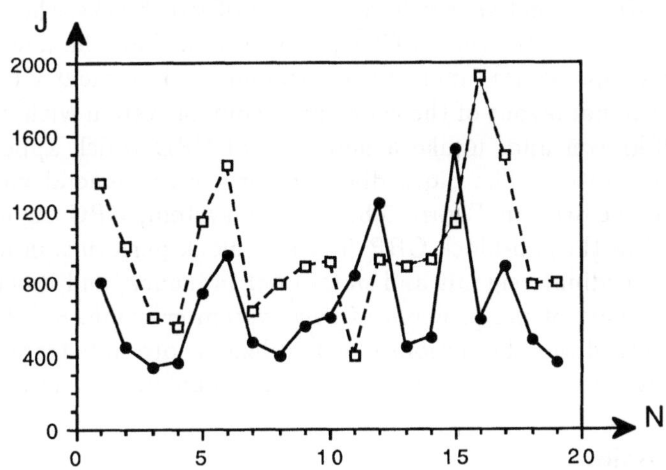

Figure 5.6. Complex criterion J values vs. trial number N for different gain values: $\eta_0 = 3.375\,\text{sec}^{-1}$ - a continuous line, $\eta_1 = 2.025\,\text{sec}^{-1}$ - a dotted line. From (Slutski and Coiffet, 1996). ©Cambridge University Press.

below line $J_{\eta_1}(N)$. On the considered interval, the average value of criterion J for η_0 is $\overline{J}_{\eta_0} = 675$, and for η_1 is $\overline{J}_{\eta_1} = 963$. This indicates that, for optimal gain value, the system quality is 30% higher than for the nearest gain value η_1.

5.3. General Organizational Principles of Parameter Control in Remote Manipulators

In spite of the remarkable increase in manual control quality, the described technical solution freezes the manipulator parameters and does not permit rapid alteration of its indices when operation requirements change. If so, another technical solution is needed to flexibly change system characteristics.

Such a solution essentially requires the HO him/herself to adjust the manipulator parameters during the control process according to changing technological requirements. The additional channel of parameter control is then introduced into the system. This channel acts simultaneously with the main coordinate channel; i.e., with the control channel of generalized manipulator coordinates. Such an approach results in the organization of semiautomatic control systems with variable parameters, unlike the better-known semiautomatic control systems with variable structure (Malone, 1973; Medvedev *et al.*, 1978). Thus, a new class of adaptive coordinate-parameter control (CPC) systems for telerobots is obtained.

An extended general block diagram of such a system is shown in Figure 5.7. The HO model shown here consists of two blocks: the traditional controller of coordinate control CCC, and, in addition, the new controller (with the new operator's function) of parameter adjustment CPA.

The traditional layout of the coordinate control system with visual feedback is well-known and, unlike a new unit of CPC which appears in the robot control system. The block diagram for a more general view of such a system can be seen in Figure 5.8.[1] In this system, CPC algorithms are implemented in the sub-block CBP. The sub-block performs, in fact, a convolution of coordinate signals and parameter influence, and its elaboration is a principal part of the problem of new system creation. Sub-block CBT performs the traditional functions of generalized coordinate transformation, and produces signals $u_1(t),\ldots,u_n(t)$ to control manipulator drives.

5.4. Discussion

The experimental investigations described above allowed us to establish the most influential telerobot parameter of control system quality. This turned out to be (as we could easily guess) the manipulator gain, whose influence on accuracy and fast-operation indices is shown above. A possible fundamental method of tuning this parameter was demonstrated. Results testify that the described method of manipulator parameter tuning improves telerobotic system performance.

However, this technical solution does not permit the rapid alteration of system quality when needed. It is therefore a worthwhile method in cases

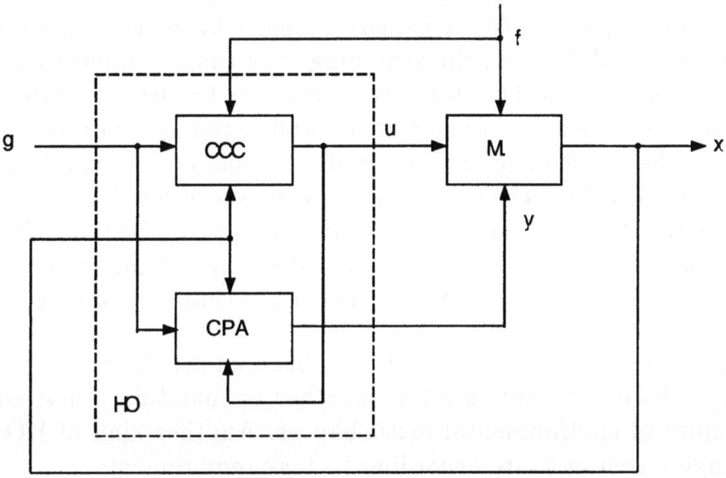

Figure 5.7. Block diagram of an adaptive remote control system. Designations: M is a controlled manipulator, CCC is a controller of coordinate control, CPA is a controller of parameter adjustment, g is an input signal; x is an output signal; f is a disturbing influence; u is a controlling coordinate signal, and y is a supplementary controlling parameter signal. From (Slutski and Coiffet, 1996). ©Cambridge University Press.

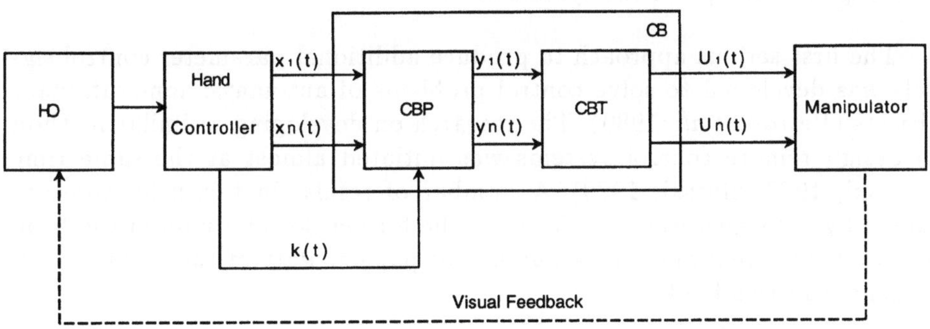

Figure 5.8. Block diagram of an adaptive telerobotic system. Designations: HC is a hand controller, CB is a control block, CBP is a sub-block of coordinate-parameter control, CBT is a sub-block of geometrical/kinematical transformations of signals. From (Slutski, 1996). ©SIRI - Associazione Italiana di Robotica, Milan, Italy.

when the teleoperator performs technological operations that do not require maximum accuracy or fast-operation.

In principle, the approach is related to gain adjustment, known in telerobotics. For instance, adjustable scaling - not only displacements from master to slave but also forces from slave to master - is mentioned in (Salcudean and

Yan, 1994). As for scaling change during HO work, a change in gain in MSSs (Saenger and Pegden, 1973) is usually realized by switching, which poses challenges for an HO. First, the switching may cause complicated dynamic processes in the system. Furthermore, it requires the HO to adapt to a new gain value, leading to loss of his/her skill and requiring HO re-education. It is clear that these factors reduce the quality of operator-manipulator system performance. Another technical solution consists of flexible adjustment of the parameter that is most influential in leading to adaptive CPC system organization. In all these cases, a physical change of manipulator gain is performed, since the final effect obtained is the changing of system scaling.

Adaptive possibilities of the HO have been known for a long time (Li *et al.*, 1966). These possibilities allow him/her to maintain a desired system state in spite of environmental disturbances. Another kind of HO activity is to change a system state according to task requirements.

The problem is, however, that such change is usually produced only with the help of some means of coordinate control, which decreases the effectiveness of adaptivity. At the same time, HO abilities permit the production of, not only the coordinate control signals group, but also additional parameter signals. We use this property of HO to enlarge his/her possibilities for adaptive changes of the system state.

The first serious approach to produce additional parameter control signals was developed to solve control problems of automatic non-stationary objects (Petrov *et al.*, 1980). The research on developing a similar method to design remote control systems was initiated almost at the same time (Slutski, 1977; Slutski, 1979). A number of results in this field were obtained by us, which were published in the former Soviet Union in the form of about 20 inventions and a number of papers. Part of these results are presented in this book.

Common ideological ground for the two approaches of CPC implementation mentioned above has not obscured a principal difference in their organization with regard to controlled object properties. The considered method allows one, in fact, "to play" with object parameters, which leads to further "play" with system performances. However, "the rules of play" are different for different applications of the method. The synthesis methods of control algorithms and technical system implementation are therefore completely different for these kinds of systems. We pass now to consider these problems for the kind of control systems that form the subject of this book and to determine "the rules of play" for semiautomatic remote systems.

References

Handlykken, M. and Turner, T. (1980) Control systems analysis and synthesis for a six degree-of-freedom universal force-reflecting hand controller, *Proc. of 19th IEEE Conf. on Decision and Control*, Albuquerque, NM, Dec. 10-12, pp. 1197-1205.

Li, Y.T., Young, L.R., and Meiry, J.L. (1966) Adaptive functions of man in vehicle control systems, in P.H. Hammond (ed.), *Theory of Self-Adaptive Control Systems*, Plenum Press, New York, pp. 43-55.

Malone, T.B. (1973) Man-machine interface for controllers and end effectors, in E. Heer (ed.), *Remotely Manned Systems*, California Institute of Technology, Pasadena, pp. 319-326.

Medvedev, V.S., Leskov, A.G., and Yushchenko, A.S. (1978) *Control Systems of Manipulation Robots*, Nauka, Moscow (in Russian).

Petrov, B.N., Zemlyakov, S.D., and Rutkovsky, V.Y. (1980) *Adaptive Coordinate-Parameter Control of Non-Stationary Objects*, Nauka Moscow (in Russian).

Saenger, E.L. and Pegden, C.D. (1973) Terminal pointer hand controller and other recent teleoperator controller concepts: Technology summary and application to Earth orbital missions, in E. Heer (ed.), *Remotely Manned Systems*, California Institute of Technology, Pasadena, pp. 327-336.

Salcudean, S.E. and Yan, J. (1994) Towards a force-reflecting motion-scaling system for microsurgery, *Proc. of IEEE Intern. Conf. on Robotics and Automation*: San-Diego, USA, pp. 2296-2301.

Slutski, L.I. (1972) On quality criteria of manipulator manual control, *Mashinovedenie* (Moscow), **no. 2**, pp. 28-33 (in Russian).

Slutski, L.I. (1977) Manipulator, USSR author's certificate # 561656, *Bulletin of Inventions* (Moscow), **no.22** (in Russian).

Slutski, L.I. (1979) Operative correction at manipulator velocity control, *Proc. of 1st Interinstitute Conf. on Robotics. Control and Sensitive Systems*, **Vol. 1**, Kaunas, pp. 36-40 (in Russian).

Slutski, L. (1995) Optimal parameter adjustment of telerobot servodrive, *Proc. of Eighteen IEEE Convention in Israel*, 7-8 March, Tel-Aviv, pp. 2.3.2.1-2.3.2.5.

Slutski, L. (1996) Adaptive telecontrol for industrial robot teaching, *Proc. of 27th Intern. Symp. on Industrial Robots*, 6-8 October, Milan, Italy, pp. 47-52.

Slutski, L. (1997) On-line telecontrol techniques based on object parameter adjustment, *Presence*, **Vol. 6 no. 3**, pp. 255-267.

Slutski, L.I. and Safontsev, E.A. (1983) Manipulator dynamics influence on remote control quality, *Theory of Automatic Control and Robotics Systems*: Collection of scientific papers, Moscow, Institute of Radiotechnique, Electronics and Automation, pp. 37-41 (in Russian).

Slutski, L.I., Safontsev, E.A., and Vitukova, G.A. (1983) Remote manipulator dynamic synthesis methods based on experimental research of control process, in *Analysis and Synthesis of Mechanisms*: Collection of scientific papers, Alma-Ata, Kazakh State University, pp. 31-37 (in Russian).

Notes

[1] The figure appeared also in (Slutski, 1997); ©1997 by the Massachusetts Institute of Technology.

CHAPTER 6

ADAPTIVE COORDINATE-PARAMETER CONTROL (CPC) OF THE REMOTE MANIPULATORS

Remote Manipulation Systems with an Additional Channel of Parameter Control

6.1. Classification of Adaptive CPC Systems

A classification of telerobot adaptive CPC systems is presented in Figure 6.1. The following classification levels should be noted here:
- control mode;
- means of parameter signal setting;
- location of parameter signal realization;
- method of parameter setting.

On the control mode, adaptive systems are divided similar to ordinary control systems (see Section 4.1).

Parameter signal setting may be realized by the following means:
- an additional input element;
- an ID of special type; or
- by sacrificing its automatic generation in the control system.

These versions will be described below in Section 6.3.

As a location of parameter influence realization note the following:
- a control system ID (a HC (see Section 6.3) or a slave arm (see Chapter 7));
- a control unit (e.g., system computer (Section 6.2)); or
- an EO that is a manipulator (Chapter 7).

The parameter may be set in the form of continuous or discrete signal. A particular case of the latter is the relay switching version. These possibilities relate to parameter signal organization and they will be proposed in Section 6.3.

All these CPC peculiarities are discussed below in detail. But the principal part in system organization is the derivation of realized-in-system algorithms and we therefore will elaborate on this problem.

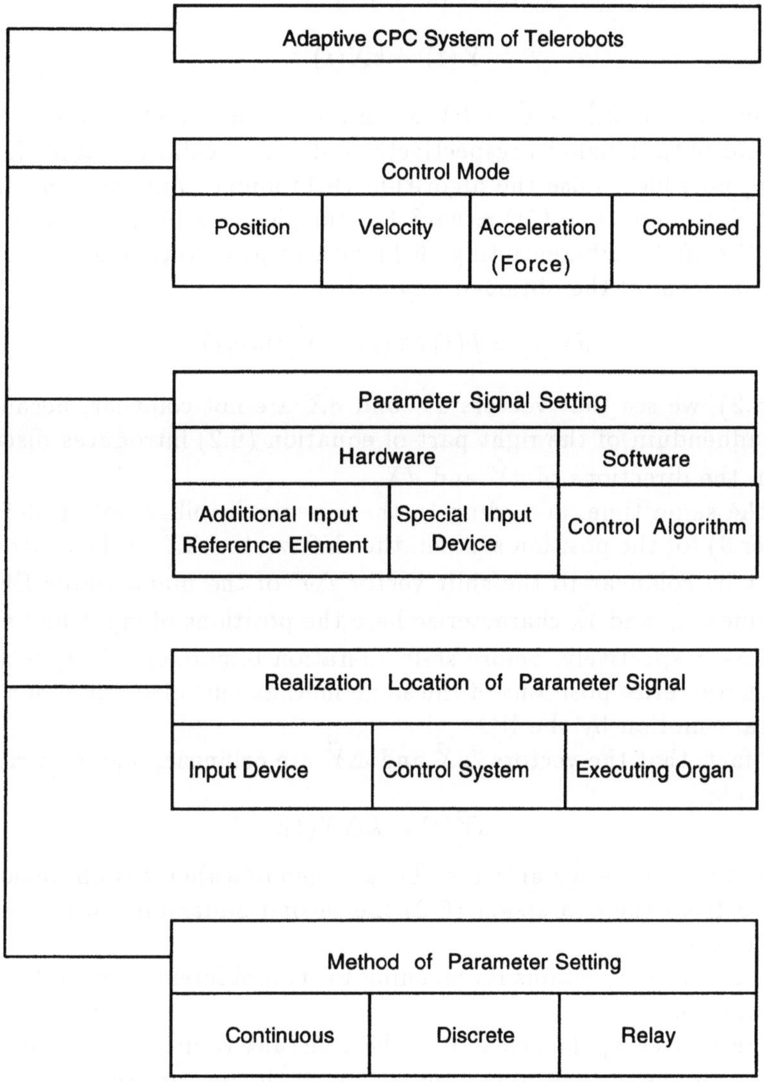

Figure 6.1. Classification of telerobot adaptive systems.

6.2. Adaptive Algorithm Synthesis

6.2.1. ALGORITHM DEVELOPMENT FOR CONTINUOUS POSITION CONTROL

When determining algorithms of the block CBP (Fig. 5.8) (Slutski, 1997b), we proceed from the assumption that in the position system under discussion, a relationship (see Subsection 4.5.1)

$$\vec{Y}(t) = k\vec{X}(t)\ , \tag{6.1}$$

is usually realized where $\vec{X}$ and $\vec{Y}$ are n-dimensional vectors of input coordinate and output signals, respectively, and k is a scalar constant. Is it, for example, possible to use the algorithm (6.1) under adaptive control; that is, when parameter $k = k(t) = var$? For this, let us find the velocity of the robot EE shift by differentiating (6.1) with respect to time and consider a differential form of the obtained expression

$$d\vec{Y}(t) = k(t)d\vec{X}(t) + \vec{X}(t)dk(t). \tag{6.2}$$

From (6.2), we see that vectors $d\vec{Y}$ and $d\vec{X}$ are not collinear, because the second addendum of the right part of equation (6.2) introduces disconcordance in the directions of $d\vec{Y}$ and $d\vec{X}$.

At the same time, according to the mnemonicability notion definition (Chapter 3) for the position system, the shift vector $\Delta\vec{X}$ of the control handle must be collinear to the shift vector $\Delta\vec{Y}$ of the manipulator EE (Fig. 6.2). Values $\vec{X}_0$ and $\vec{Y}_0$ characterize here the positions of input and executing organs, respectively, before shift operation fulfillment; thus, vectors $\vec{X}$ and $\vec{Y}$ characterize positions of the same mechanisms after fulfillment of an elementary motion by the HO.

The fact, that the vectors $\Delta\vec{X}$ and $\Delta\vec{Y}$ are collinear, may be written as the formula

$$\Delta\vec{Y}(t) = k\Delta\vec{X}(t), \tag{6.3}$$

where k is a positive scalar value. Thus, mnemonicability is obtained while excluding from the expression (6.2) the second addendum, which violates mnemonicability.

Expression (6.3) retains its meaning for two different cases of the scalar k presentation.

In the first case, the scalar k is the constant value: relation (6.3) then characterizes the ordinary position control with constant scaling according to algorithm (6.1).

This scalar may also be a variable value $k(t) = var$, which corresponds to the CPC case. For this version, it is possible to perform some transformations of expression (6.3) in order to obtain different forms of adaptive algorithms.

Algebraic Algorithm Form

Let us consider two vector relations (Fig. 6.2) for the manipulator EO

$$\vec{Y} = \vec{Y}_0 + \Delta\vec{Y}, \tag{6.4}$$

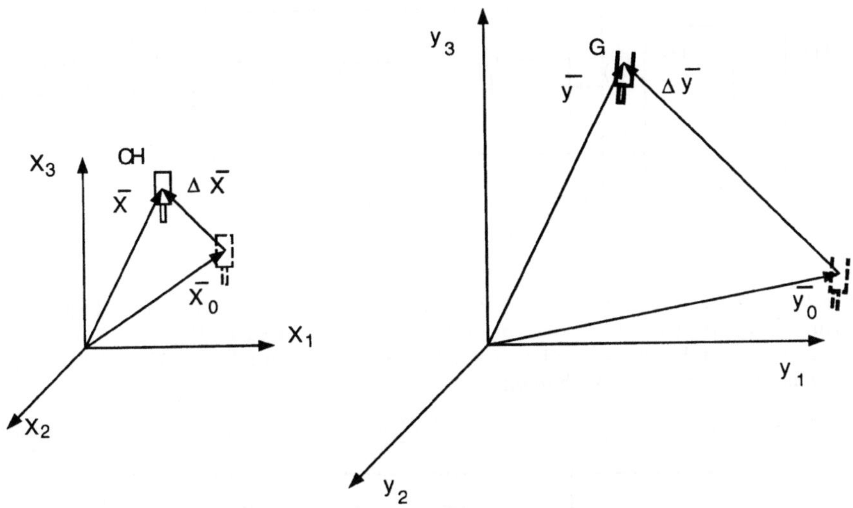

Figure 6.2. Toward the mnemonicability notion definition. Correlation between shifts of the input and executing organs of a manipulator. Designations: CH is a control handle, G is a robot EE. From (Slutski, 1997b). ©1997 by the Massachusetts Institute of Technology.

and for a HC

$$\vec{X} = \vec{X}_0 + \Delta\vec{X}. \tag{6.5}$$

If we link equations (6.4) and (6.5) with expression (6.3), we obtain (Slutski and Coiffet, 1996)

$$\vec{Y}(t) = \vec{Y}_0 + k(t)[\vec{X}(t) - \vec{X}_0]. \tag{6.6}$$

To make a device that will implement the algorithm is somewhat complicated, because it must contain memory and requires that the current state of the vectors $\vec{X}_0$ and $\vec{Y}_0$ be measured and remembered. Algorithm (6.6) also requires that an interrogation of the control signals be executed with sufficiently high frequency to ensure necessary control accuracy. Hence it is worthwhile to proceed to another form of the adaptive algorithm.

Integral Algorithm Form

If we transform formula (6.3) using the differentials, we arrive at the simple transformations

$$\vec{Y}(t) = \int_0^t k(t)\dot{\vec{X}}(t)dt, \quad \vec{Y}(0) = 0. \tag{6.7}$$

When $k(t) = k = const$, we obtain from (6.7) the correlation for conventional coordinate control (6.1).

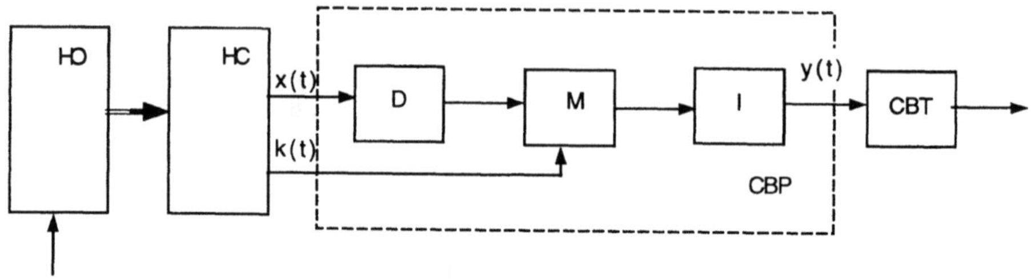

Figure 6.3. Adaptive device layout for position control (one controlled channel). The following designations are labeled (in addition to those used in Figure 5.7): I is an integrator; D is a differentiator; M is a multiplier. From (Slutski, 1997b). ©1997 by the Massachusetts Institute of Technology.

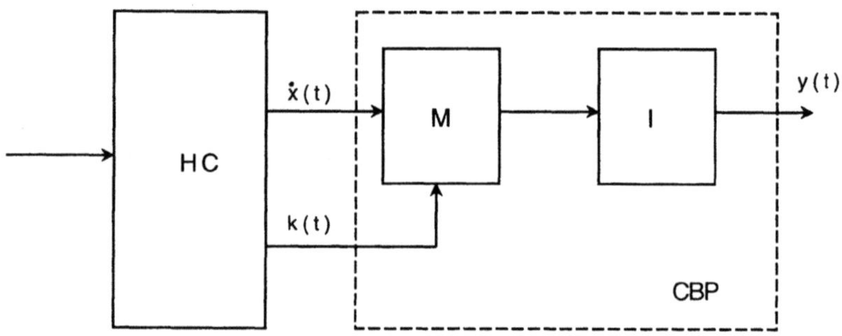

Figure 6.4. Adaptive device layout for position control with velocity sensor use (one controlled channel). I is an integrator, and M is a multiplier. From (Slutski, 1997b). ©1997 by the Massachusetts Institute of Technology.

The control device layout (Fig. 6.3) corresponds to the algorithm (6.7). In this layout, integrator I acts also as the memory. Lastly, this device structure may be further simplified if the velocity transducers are used as the sensors. Tachometer sensors, for example, must be placed for this in HC joints to measure relative velocities of HC link shifts. The need for differentiator can thus be avoided, and, as result, the obtained device will be even simpler (Fig. 6.4).

6.2.2. A SYSTEM WITH STEP SETTING OF THE PARAMETER SIGNAL

As described in 6.2.1, the systems were implemented by realizing the principle of continuous parameter change. Another approach is connected with step setting of the parameter signal. A sample of such a system is shown in Figure 6.5.[1] Button (2) switches here work modes, and amplifier (6) has

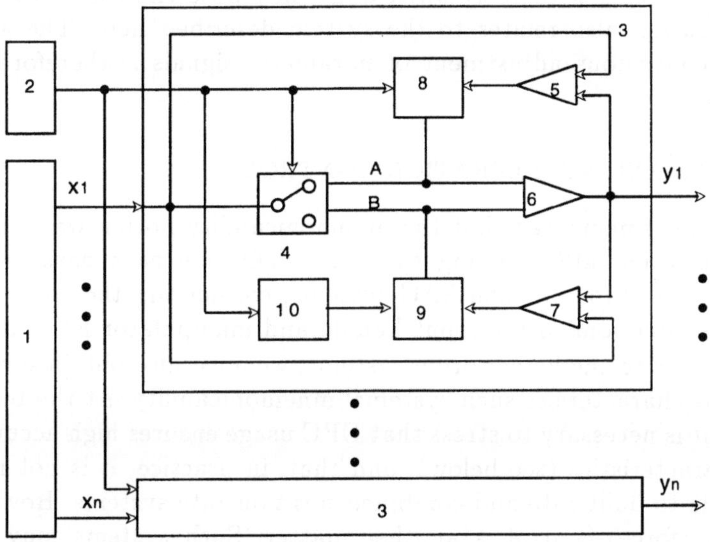

Figure 6.5. **Layout of adaptive position control system with step setting of parameter signal. Designations: 1 is a HC; 2 is a button; 3 are blocks for control signal transformation; 4 is a controlled switch; 5, 6, and 7 are summing operational amplifiers; 8 and 9 are blocks of memory; and 10 is a logic element "NO." From (Slutski, 1997a). ©1997 by the American Nuclear Society, La Grande Park, Illinois.**

two inverted inputs, as well as a maximum gain k_1, which corresponds to input A, and a minimum gain k_2, corresponding to input B. Operational amplifiers (5) and (7) have non-inverted inputs, and an input signal x is enhanced in the amplifier (5) with a gain k_2, and in the amplifier (7) with a gain k_1.

This device operates in two modes: one with maximum gain k_1 and one with minimum gain k_2. Blocks of memory (8 and 9) ensure switching from mode to mode without jumps, which are considerable transient processes.

When the system works with gain k_1, the switch (4) is in position A. The memory (8) then records the amplifier (5) input signal and the memory (9) restores a previously recorded signal. When the system passes to the operating mode with gain k_2, the switch (4) is switched to position B. The memory block (8) passes then to a restore mode, and the switching moment signal recorded from the memory (8) enters the amplifier (6). Here, it is summed with the signal from the HC. Thus, a smooth passage from one mode to another occurs and the system begins to work with the low gain k_2.

This system is, in fact, an analog of the above described combined position-rate system (Subsection 4.5.1). It has already been pointed out that such systems with relay switchings may pose challenges to the HO. This conclusion also relates to the system described here. The approach with the continuous adjustment of parameter signals is therefore further developed.

6.2.3. RATE AND ACCELERATION CONTROL

It should be remembered that the mnemonicability notion was introduced only for position MSSs. Using the main definition as a basis, we wrote correlation (6.3) for position SAC systems, considering the coincidence of the vector directions of the input handle and manipulator EE shifts.

Passing on to the higher order systems, we must also obtain some correlation that characterizes such systems' mnemonicability. At the beginning, however, it is necessary to stress that CPC usage ensures high accuracy and speed characteristics (see below), and that, in practice, it is not normally worthwhile to built rate and combined position-rate systems. However, acceleration (force) control is another matter. Such systems may be very widely used in practice, for example, in combination with position control - a combination which is likely to be very useful. Therefore, it is important to develop control algorithms for high order systems.

Rate Control
By analogy with position control, and using the induction principle, we can define the mnemonicability for a rate system as the coincidence of the directions of the shift vector of the ID and the velocity increment vector of the manipulator EE. The correlation may be written as

$$\Delta\dot{\vec{Y}}(t) = k_R(t)\Delta\vec{X}(t).$$

After its transformation, we obtain the adaptive rate control algorithm in remote systems (Slutski, 1979):

$$\vec{Y}(t) = \int_0^t k_R(t)\vec{X}(t)dt \text{ , } \vec{Y}(0) = 0. \tag{6.8}$$

Acceleration Control
By extending the above analogy, in the case of acceleration control, mnemonicability is characterized by the correlation between directions of the ID shift vector and the acceleration increment vector of EE. We may then write the following condition for acceleration control:

$$\Delta\ddot{\vec{Y}}(t) = k_A(t)\Delta\vec{X}(t) \text{ .} \tag{6.9}$$

After transformation of expression (6.9), the algorithm is obtained (Slutski, 1995)

$$\vec{\ddot{Y}}(t) = \int_0^t k_A(t)\vec{\dot{X}}(t)dt \; , \; \vec{\ddot{Y}}(0) = 0. \tag{6.10}$$

6.2.4. CONTROL BY MEANS OF HAND CONTROLLER ACCELERATION

An additional possibility for remote control is opened when the initial expression (6.7) is differentiated twice with respect to time, which results in the correlation

$$\vec{\ddot{Y}} = \dot{k}(t)\vec{\dot{X}}(t) + k(t)\vec{\ddot{X}}(t) \; . \tag{6.11}$$

By integrating this twice, we obtain the algorithm

$$\vec{Y}(t) = \int_0^t \int_0^t [\dot{k}(t) \int_0^t \vec{\ddot{X}}(t)dt + k(t)\vec{\ddot{X}}(t)]dtdt \; . \tag{6.12}$$

Here, the control is set by altering the acceleration of the input handle and not by means of handle position or velocity changing. We must use acceleration gauges as sensors that give original properties to the HC. In this case, the handle moves freely in the operator's hand and is not connected to the base.

A schematic setup of such a HC is shown in Figure 6.6. It is clear that such a control method leads to special control technology and requires corresponding training of the HO. However, the positive advantages of such an approach may be decisive.

These examples are related to position (6.12) and acceleration (6.11) manipulator control, but it is clear that it is also possible to control manipulator gripper velocity. By differentiating (6.11), we obtain the following algorithm:

$$\vec{\dot{Y}}(t) = \int_0^t [\dot{k}(t) \int_0^t \vec{\ddot{X}}(t)dt + k(t)\vec{\ddot{X}}(t)]dt \; .$$

The HC example shown in Figure 6.6 is primarily intended to control manipulator transportation motions. Realization of its orientation motions requires the use of either additional sensors or of additional program means. Ordinary pushbuttons for corresponding axis may be used as additional sensors. For practical control purposes, these are usually sufficient, since these motions do not normally demand very high accuracy or speed.

If necessary, however, handle sensors (Fig. 6.6) may be used. Signals from sensors A_1, A_2, and A_3 are produced not only during handle linear shift, but also during its rotations about the axis x, y, z. In fact, while rotating the handle about the x axis, the sensors A_2 and A_3 produce corresponding signals, depending on rotation angle speed. In a similar manner,

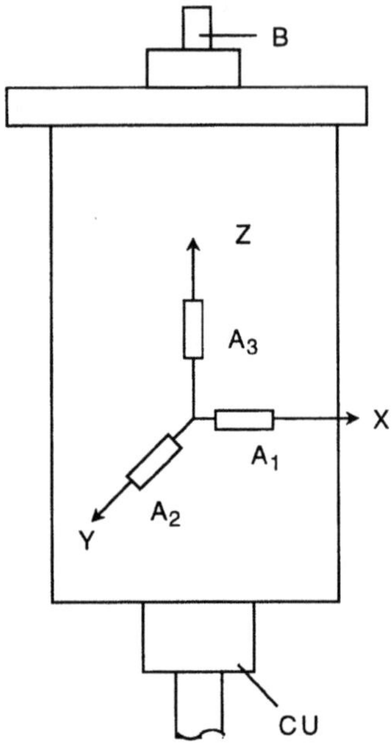

Figure 6.6. Layout of a HC based on accelerometer sensors. Three accelerometers, A_1, A_2, A_3, are placed here according to three coordinate axes. The device also has a button (B) for robot gripper control, and a connection unit (CU) that transfers information into the control system. From (Slutski, 1997b). ©1997 by the Massachusetts Institute of Technology.

when the handle rotates about the y axis, sensors A_1 and A_2 are active, and when the handle rotates about the z axis, sensors A_1 and A_3 are active. These facts may be used to organize a control system for gripper orientation. There is one difficulty here in determining rotation direction: the sensor signals are not direction dependent. The solution of this problem requires placement of additional simple switchers on the HC.

6.2.5. AUTOMATIC PARAMETER ADJUSTMENT

In some forms of operations of telerobotic systems, the flow of information reaching the HO that he/she must organize using the HC exceeds his/her information processing capabilities. In such a case, it is clearly not practical to burden the HO with an additional parameter signal-producing function. This work should therefore be handled by a special automatic device. A simple logical process may serve as the basis of its operation.

Highly precise manipulation is usually realized by the HO using slow, "creep" velocities in all manipulator joints. At the same time, adaptive algorithms are obtained assuming that minimum manipulator gain corresponds to maximum positioning accuracy, and that maximum gain corresponds to maximum transportation speed. We may then relate to each of the other shift velocities $\dot{x}_i, i = 1, ..., n$ (n is a number of controlled channels) of the HC axis and manipulator control system gain. For one control channel, the dependence is

$$k_i(t) = F(|\dot{x}_i|),$$

which is more simply realized with the help of a proportional correlation

$$k_i(t) = k_{min} + \alpha|\dot{x}_i|, \alpha = const, k_{min} = const. \tag{6.13}$$

Here, the value k_{min} ensures maximum accuracy of teleoperation.

However, by using the dependences (6.13), we may obtain inequality of different channel gains and, accordingly, mnemonicability deterioration. To keep a mnemonicability level, it is necessary to ensure the equality of all the gains $k_i (i = 1...n)$.

We may achieve this by using expression (6.13) in the form

$$k_i(t) = k_{min} + \alpha|\dot{x}_i^0(t)| \tag{6.14}$$

and by choosing values $\dot{x}_i^0$ as one of following kinds:

$$|\dot{x}_i^0(t)| = max|\dot{x}_i(t)|, i = 1...n, \tag{6.15}$$

$$|\dot{x}_i^0(t)| = \frac{1}{n}\sum_{i=1}^{n}|\dot{x}_i(t)|. \tag{6.16}$$

Version (6.15) uses energetic robot possibilities more fully. Maximum values of joint velocities are chosen here based on the considerations discussed below.

For precise task performance, low joint velocities usually occur in all manipulator joints. If a manipulator executes a fast operation, it may be performed, depending on manipulator structure, with high motion velocity in one (or some) controlled joint(s) and with low velocity in other joints. But, in this situation, the shift velocity in the first group of joints should be the maximum that the algorithm (6.14) ensures.

Based on dependences (6.7) and (6.14), a new class of adaptive algorithms is obtained (Slutski, 1995). The gain adjustment is realized automatically, according, for instance, to the expression

$$\vec{Y}(t) = \int_0^t [k_{min} + \alpha|\vec{X}^0(t)|]\vec{X}(t)dt \ , \tag{6.17}$$

where the vector $|\vec{X}^0(t)|$ components may be determined by the expression (6.15).

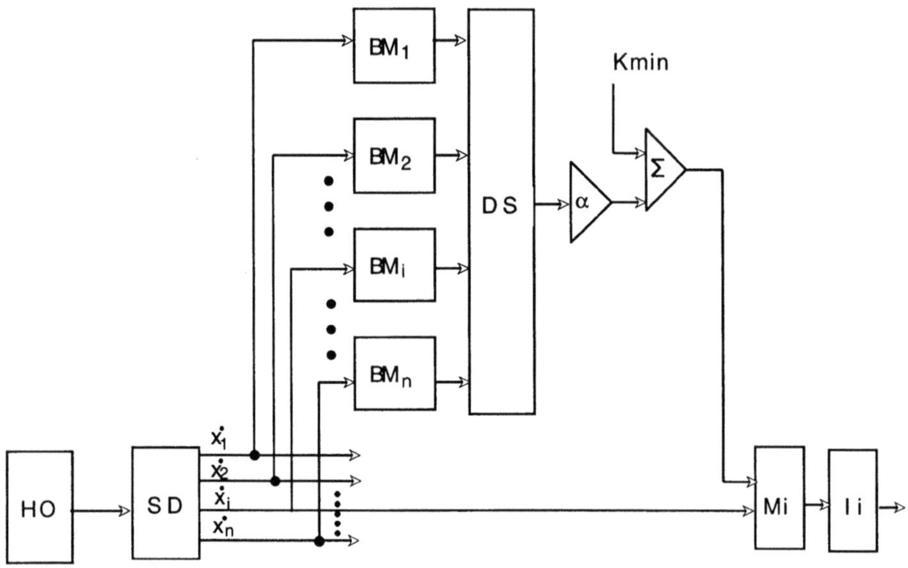

Figure 6.7. Block diagram of the control device with automatic parameter adjustment. The following new blocks are used here: $BM_1 - BM_n$ are blocks of module acquirement, where n is the number of controlled channels, DS is a discriminator of the maximum velocity $max|\dot{x}_i(t)|$, $\sum$ is a summing amplifier, M_i is a multiplier, and I_i is an integrator of the i-th channel. From (Slutski, 1997b). ©1997 by the Massachusetts Institute of Technology.

A special control device, the block-diagram of which is shown in Figure 6.7, corresponds to this version of a CPC system.

We see that, in this case, control of the parameters of the coordinate controller takes place, but not quite the control of object parameters. Note that the automatic parameter adjustment may be realized, not only for the position control (6.17), but also for other types of systems.

When using algorithm (6.17), there is a certain restriction on the gain of one control channel:

$$k(t) = k_{min} + \alpha|\dot{x}(t)| \leq k_{max}, \tag{6.18}$$

where k_{max} is the maximum gain value from the viewpoint of the stability of the operator-manipulator system. The k_{min} value is set from a condition of maximum accuracy achievement. Estimation of the value $|\dot{x}(t)|$ may be obtained based on ergonomic data about maximum velocities V_{HOmax} that a human arm develops. Therefore, an estimate from above of the constant parameter α may be obtained from the expression (6.18)

$$\alpha \leq \frac{k_{max} - k_{min}}{V_{HOmax}}. \tag{6.19}$$

There is an additional consideration for choosing parameter α. According to our experimental data obtained during adaptive system simulations (which will be described in Chapter 8), when parameter α is increased, a decrease of control process velocity takes place which is obtained by the HO decreasing the shift amplitude of the joystick handle. The dependence of this index on α has an inverse character. This suggests the means required to reduce overall dimensions of the input device and the HO motion amplitude in order to reduce his/her fatigue.

In practice, parameter α may be chosen first in the design stage, in order to obtain, for instance, minimum amplitude of the HC handle shift and small dimensions of a hand controller. It may also be tuned during system operation to reach system optimal characteristics; that is, either small human arm shifts or minimum operation time, as needed.

6.3. HCs of Adaptive Systems

HCs that are in use in adaptive systems are, in fact, HCs of SAC systems, since systems with a parameter adjustment are a variety of semiautomatic systems. Therefore, according to the classification in Section 6.1, even ordinary HCs of semiautomatic systems may be used for adaptive control. However, in many cases, an adaptive system application requires a specific design of HC, and this design is tightly connected with the type of control algorithm used. More frequently, HCs have to be provided by additional input reference elements for parameter signal setting. Lastly, special HCs may be organized that produce a parameter signal only because of specific constructions; that is, without placement of an additional input reference element. Therefore, we are describing in this chapter the specific design of HCs for adaptive systems. Common demands for all these varieties of HCs are their small dimensions and small working area of the HO's hand.

6.3.1. CONVENTIONAL HAND CONTROLLER USE FOR ADAPTIVE SYSTEMS

In adaptive systems, conventional HCs may be used only in the case of automatic parameter signal adjustment (see Subsection 6.2.5). These conventional HCs and joysticks are described in Section 4.2. We present here a special type of HCs that is intermediate between those based on the use of lower kinematic pairs and isometric HCs whose feature is the use of higher kinematic pairs.

One example of such HCs for telerobot control (Figs. 6.8 and 6.9)[2] consists of a base (1) and a handle (2) that is rigidly bound with the fingers (3) and which makes a cross-bar with them. The handle (2) is established on the base through an elastic suspension built in the form of flat springs

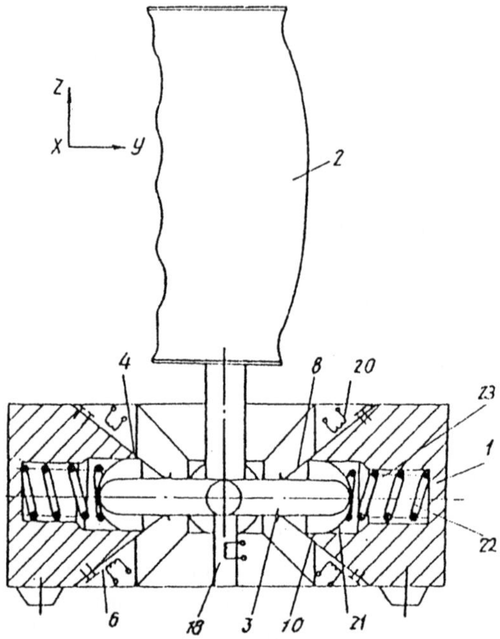

Figure 6.8. Telerobot 6-axis HC. The following principal designations are used: 1 is a base; 2 is a handle; 3 are fingers; 4-19 are flat springs; and 20 are strain gauges. From (Slutski, 1996). ©SIRI.

(4-19) with deformation sensors (for example, strain gauges) (20). Each of the fingers (3) interacts with the free ends of four springs corresponding to it. Each finger also interacts with a special mechanism that returns the finger to its initial position. This mechanism is made in the form of a spring elastic stop (21) with a spherical hollow on its operational surface. Stop (21) is backed by a spring and is situated in a hole (23) of the base (1).

While controlling this device, the HO manipulates the handle (2) by rotating and inclining it. As this takes place, the fingers (3) deform the flat springs (4-19) according to control handle motions. The deformation degree of the springs is measured by the sensors (20), signals of which enter a robot control device. The stop (21) create additional counteraction the HO efforts, and allows considerable effort to be produced by rather minor deflections of the handle (2). Based on this construction, variable efforts of resistance may be implemented.

This HC permits the control of the six DOF of the robot: three transportation and three orientation motions.

If it is difficult for the HO to control all robot DOFs with a single HC,

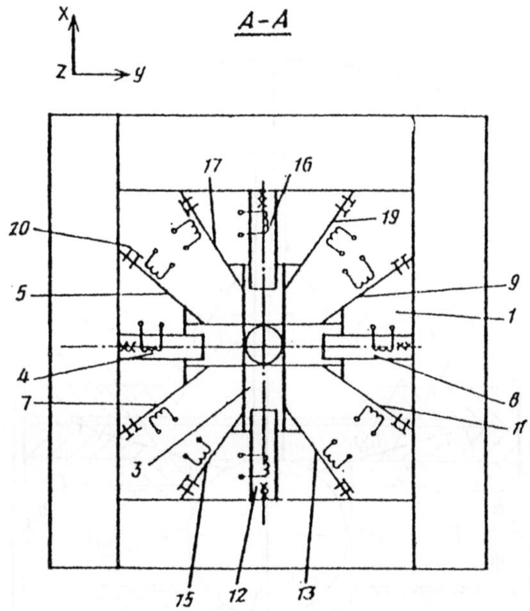

Figure 6.9. A section A-A through the HC of Figure 6.8. The designations are the same as those of Figure 6.8.

another approach to remote control is to use HCs with a smaller number of axes (see Section 4.2).

Such a HC (Figs. 6.10 and 6.11) contains a handle (1), a base (2) that is connected with flexible plates (3-8), and a pivot (9) that rigidly connects the handle (1) and a ball (10). The ball (10) has fingers (11) situated perpendicularly to the pivot (9) axis and symmetrically about the ball (10) center. The ball (10) is housed in a cylinder (12) where flexible plates (14 and 15) are placed. One end of plates (14 and 15) are rigidly bound with the cylinder, while the other end makes contact with the fingers (11). Position sensors (16-23) are placed on the plates (3-8, 14, 15), since these sensors may be used with both strain gauges and microswitches. In addition, a push-button (24) to open-close the manipulator gripper is placed on the handle (1).

This device works as follows. While moving the handle (1) in a vertical direction upward or downward, plates (5 or 6), respectively, are bent and sensors (18 or 19) are activated. While deflecting the handle (1) in the drought plane to the right or the left, plates (3 or 4) are bent and sensors (16 or 17) come into action. While deflecting the handle in the plane normal to the drought plane, plates (7 or 8) are bent and sensors (20 or 21)

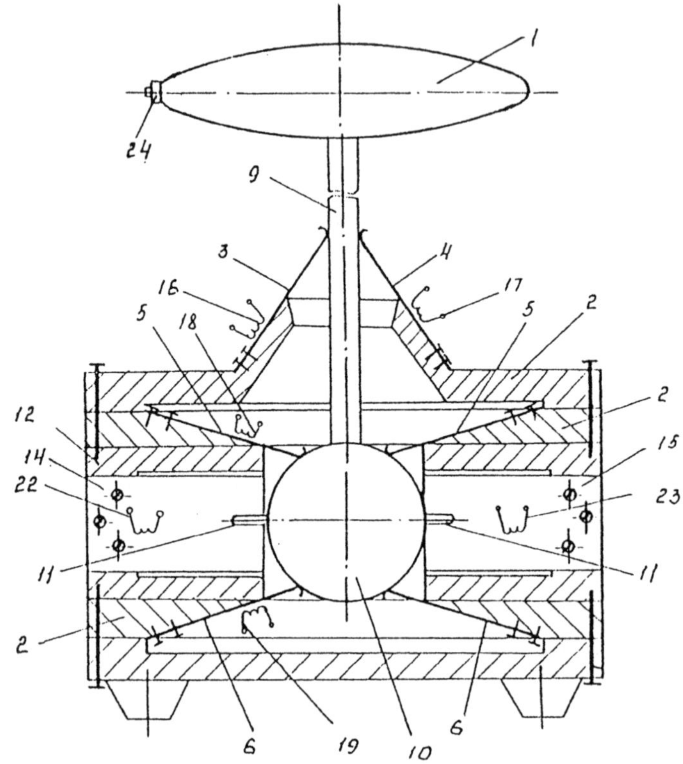

Figure 6.10. Telerobot HC with a smaller number of axis. Principal designations are: 1 is a handle; 2 is a base; 3-8, 14, and 15 are flexible plates; 10 is a ball; 11 are fingers; and 16-23 are strain gauges. From (Slutski, 1996). ©SIRI.

are activated. Further, when rotating the handle about a pivot (9) axis, plates (14 or 15) are bent and sensors (22 or 23) are activated. The sensors' (16-23) signals after transformations in the control system come to the corresponding servodrives of the robot.

This HC (Fig. 6.12) is easily controlled by the HO because the flexible plates of different axes are situated independently of one another.

This type of HC does not relate to the class of isometric HCs. A contact takes place here between the finger and the flat spring, and the contact occurs theoretically in a point. Therefore, higher kinematic pairs are in use in this mechanism, that ensures a number of original features, the main one being that these HCs are characterized by the compactness of the isometrical HCs, but not by their cross coupling of different control axes.

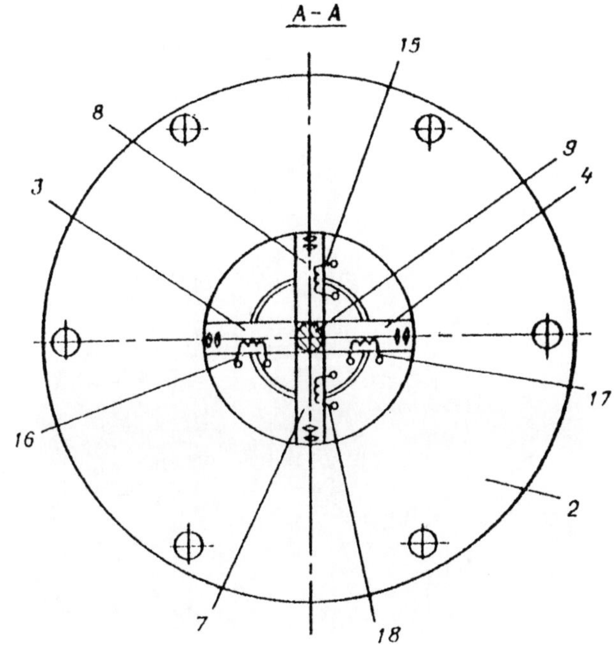

Figure 6.11. A section A-A through the HC of Figure 6.10. The designations are the same as those of Figure 6.10.

6.3.2. ADDITIONAL CONTROL ELEMENT TO SET PARAMETER SIGNAL

In spite of the possibility of using conventional HCs for adaptive control of telerobots, the HCs of the adaptive systems usually differ from those used in conventional systems in that they set up both the coordinate signals and the parameter signal. This ensures more flexible control.

The simplest case of parameter signal setting is when only a relay - that is, a step change - of the signal is introduced into the control system. In this case, it is necessary to use an ordinary two-polar switch, which is placed on the control handle. Such an input reference element is used, for example, in the system described in Subsection 6.2.2 (Fig. 6.5).

An example of a HC with a continuous setting of the parameter signal is presented in Figure 6.13 (Slutski, 1994).

Coordinate control signals are produced here as in known systems of SAC. The HO moves the handle (H) and its transferences are fixed by sensors (S-X, S-Y, S-Z). These sensors set manipulator arm spatial motions. In this example, the HC also has sensors (S-r and S-h), which set motions of rotation and clamping of the manipulator gripper, and incorporates a

Figure 6.12. Telerobot control terminal with described HC. From (Slutski, 1996). ©SIRI.[1]

special reference input element for setting the parameter signal. The latter element works when the HO squeezes the elastic handle (H) and the stop (S) presses on the elastic plate (P). The plate deforms, and sensor (S-k) gives the control parameter signal that proportionally connects the handle squeezing force F and the manipulator gain k:

$$F(t) = a \cdot k(t), \; a = const.$$

A parameter signal may also be set by using an ordinary potentiometer sensor connected with a lever (Fig. 6.14), which is pressed by the HO's fingers. The lever can be geared to the potentiometer in order to increase parameter signal range. In the photograph in Figure 6.15, a control handle is shown with a parameter input reference element where the lever is connected with the linear potentiometer.

During the control process, the operator continuously introduces into the system the maximum gain value when executing transportation movements, and the minimum gain value when executing operations requiring high accuracy. This procedure permits the design of very simple and effective telerobots with flexible control.

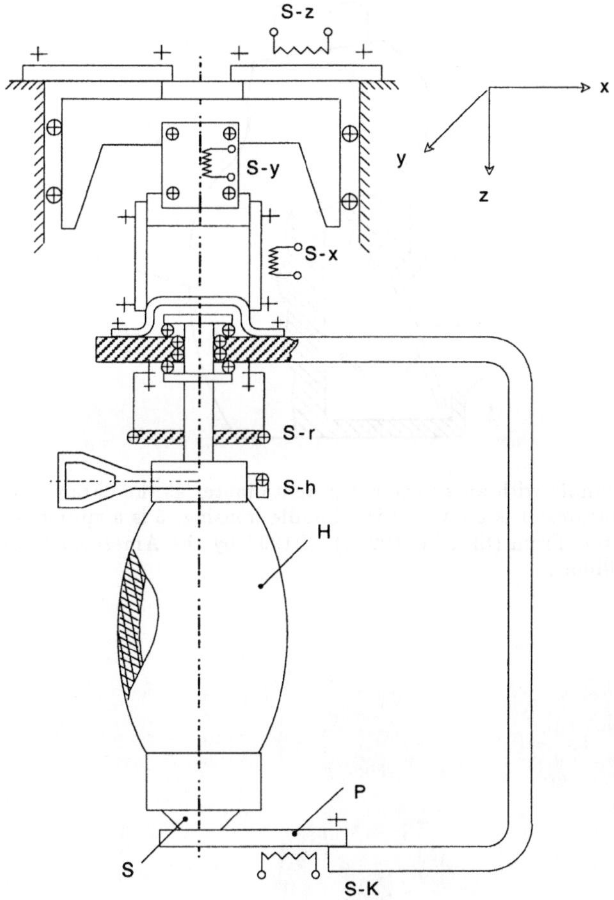

Figure 6.13. HC with special parameter input reference element based on a strain gauge. Designations: H is the elastic handle; S-X, S-Y, S-Z, and S-K are strain gauges; S-r and S-h are potentiometer gauges; P is the elastic plate; and S is the stop. From (Slutski, 1997b). ©1997 by the Massachusetts Institute of Technology.

The above examples have concerned the continuous introduction of the control signals into a telerobotic system, but a discrete setting of the parameter signal is more suitable, for example, for digital control systems, which also permits one to easily increase a changing range of parameter signals. This approach was used by us for the IR control system, and the results are described below in Subsection 6.4.2.

In the example shown in Figure 6.16, the microswitches (1-5) are placed in the form of a line. The input reference element lever (5) is rotated around an axis situated near an extreme microswitch. If one presses this lever

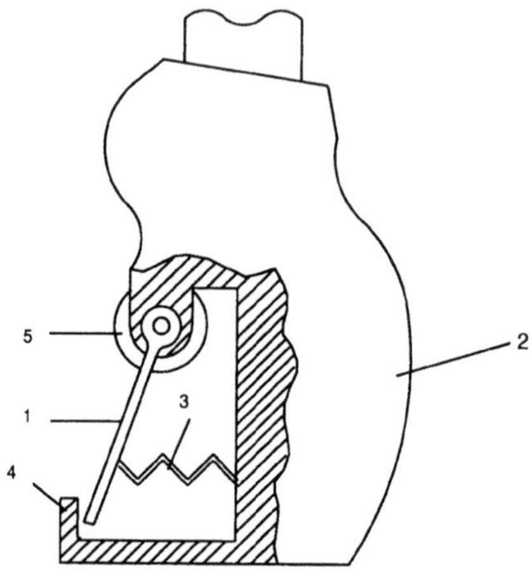

Figure 6.14. Handle with an additional potentiometer as the parameter input reference element. Designations: 1 is a lever, 2 is a handle housing, 3 is a spring, 4 is a stop, and 5 is a potentiometer. From (Slutski, 1997a). ©1997 by the American Nuclear Society, La Grande Park, Illinois.[1]

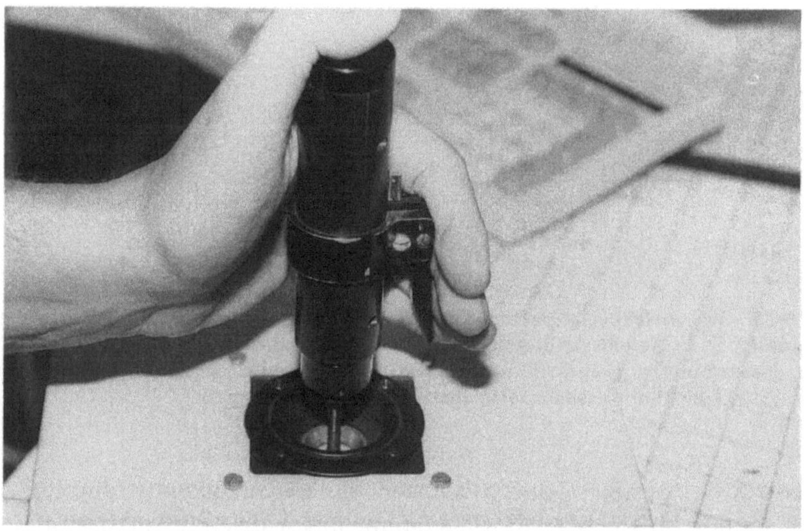

Figure 6.15. A picture of the handle with a linear potentiometer as the parameter input reference element. From (Slutski, 1997a). ©1997 by the American Nuclear Society, La Grande Park, Illinois.

smoothly, the microswitches are closed sequentially.

In our HC design (Subsection 6.4.2), this approach was implemented as follows. While releasing the lever (5) (Fig. 6.16), 1/32 of the full signal

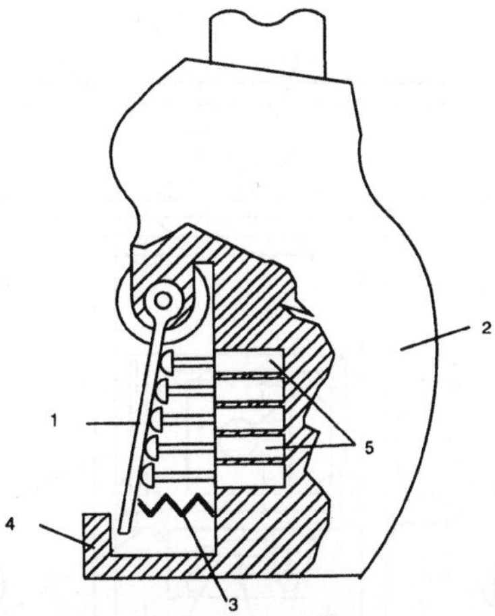

Figure 6.16. Handle with a microswitch set for discrete setting of the parameter. Designations: 1 is a lever, 2 is a handle housing, 3 is a spring, 4 is a stop, and 5 are microswitches. From (Slutski, 1997a). ©1997 by the American Nuclear Society, La Grande Park, Illinois.[1]

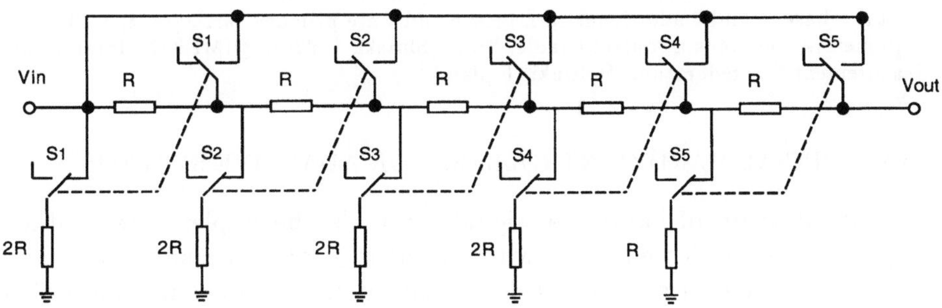

Figure 6.17. Electronic circuit to implement the parameter signal ($S_1 - S_5$ are contacts of microswitches shown in Figure 6.16). From (Slutski, 1997a). ©1997 by the American Nuclear Society, La Grande Park, Illinois.[1]

passes through each channel of the system. When the first microswitch is pressed, 1/16 of the signal passes, and then 1/8, 1/4, 1/2, and so on, until the full signal is received. Thus, the relative change range of gain runs from 1 to 1/32. The electronic circuit generating this output, called the R-2R ladder, uses a set of resistors of only two values (Fig. 6.17) (Horowitz and Hill, 1989). According to this principle of modulator creation, not only linear, but also any other relationships, may be established if needed.

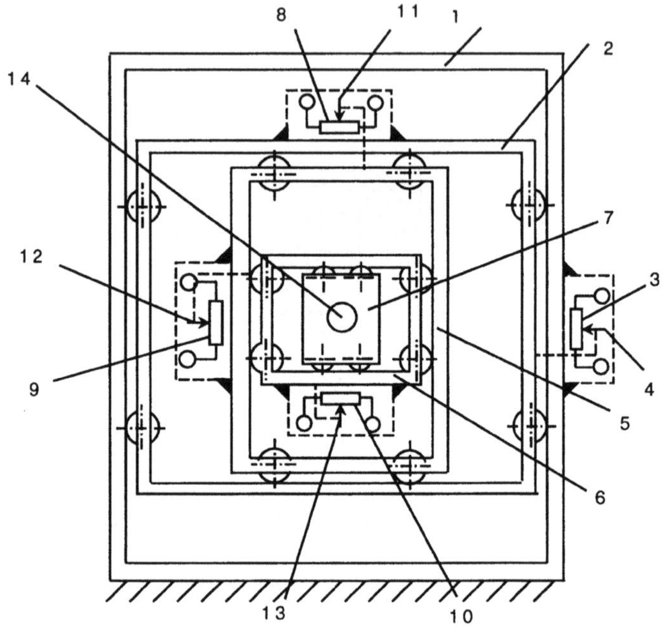

Figure 6.18. Two-dimensional HC with two pairs of carriages. Principal designations are: 1 is a base; 2 and 5 additional carriages; 6 and 7 are primary carriages; 3, 8-10 are potentiometers; and 14 is a control handle. From (Slutski, 1997c). ©IMEKO - International Measurement Confederation, Helsinki, Finland.[1]

6.3.3. SPECIAL HAND CONTROLLERS FOR ADAPTIVE SYSTEMS

The introduction of parameter signals may also be implemented without a special setting element, but with the help of an especially constructed HC. An example of such a 2D device layout is presented in Figure 6.18. It includes a base (1) with guides for transference of a first additional carriage (2). The potentiometer sensor (3) is placed on the base (1), and its movable contact (4) is connected to the additional carriage (2). This carriage contains guides for the transference of the second additional carriage (5), which has guides for the transference of the first primary carriage (6), which, in turn, has guides for the transference of the second primary carriage (7). Potentiometer sensors (8, 9, 10) of linear shifts are mounted on the guides of the carriages (2, 5, 6). The movable contacts (11, 12, 13) are connected, respectively, to the carriages (5, 6, 7). Control handle (14) is placed on the primary carriage (7). The additional carriages (2 and 5) in this construction are slightly braked. This may be accomplished, for example, by using springs, which are not shown in the Figure. Note that any orthogonal carriages may, in principle, be braked.

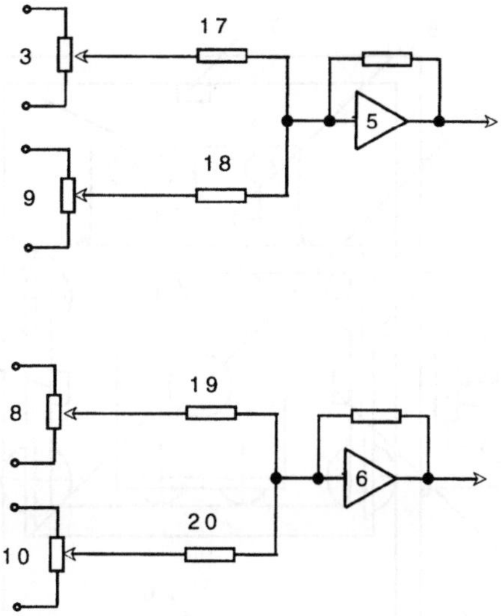

Figure 6.19. Computing device for the HC of Fig. 6.18. From (Slutski, 1997c). ©IMEKO.[1]

A computing device for this HC is described here in an analog implementation, although it may be realized, of course, in a digital way. In this computing device (Fig. 6.19), output signals of the sensors (3 and 9) of the same coordinate enter an operational amplifier (15) (inputs (16 and 17)). In a similar manner, the sensors (8 and 10) of another coordinate are connected to the inputs (19 and 20) of the operational amplifier (18). As this occurs, the outputs of the sensors (3 and 8) of the braked carriages (2 and 5) are connected with the amplifier inputs (16 and 19) with maximum gains, and the outputs of the sensors (9 and 10) of the carriages (6 and 7) are connected to the amplifier inputs (17 and 20) with minimum gains.

With this device, the HO acts on the control handle (14). As this occurs, the primary carriages (6 and 7) move because the additional carriages are braked. The signals from linear sensors (9 and 10) enter the amplifiers' inputs (17 and 20) with minimum gains. The remote manipulator then makes small transferences that ensure high accuracy of the task being performed.

When the primary carriages (6 and 7) reach extreme positions and stop because of stress, the additional carriages (2 and 5) begin to move. Signals from the sensors (3 and 8) then enter the amplifier inputs (16 and 19) with maximum gains, and they are summed up there with signals from the sensors (9 and 10), respectively. The manipulator makes large motions with a high transportation speed. If the need for accurate control occurs, the HO removes the primary carriages (6 and 7) from the stresses and works with

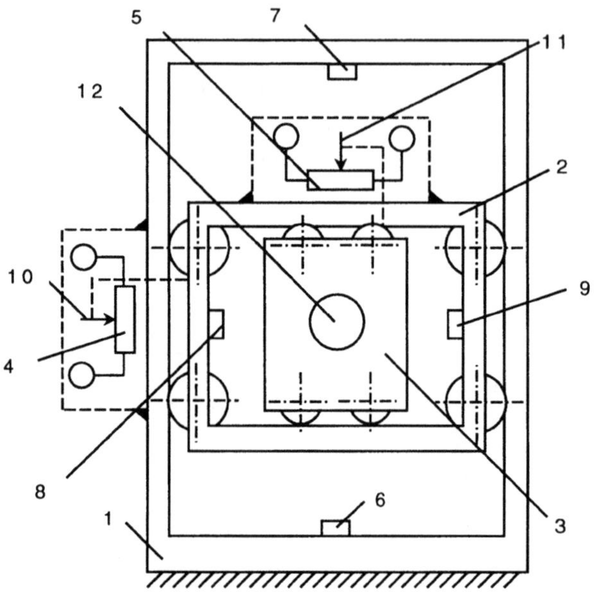

Figure 6.20. Two-dimensional HC with limit switches. Main designations are: 1 is a base; 2 and 3 are carriages; 4 and 5 are potentiometers; 6-9 are limit switches; and 12 is a control handle. From (Slutski, 1997c). ©IMEKO.[1]

them. When this device is used in a position mode, the control algorithm is

$$x_C = k_1 x_1 + k_2 x_2, \quad y_C = k_1 y_1 + k_2 y_2,$$

where x_C, y_C are the vector components of robot gripper shift; x_1, y_1 are the primary carriage shifts; x_2, y_2 are additional carriage shifts; k_1, k_2 are gains; and $k_1 << k_2$.

Another modification of this device ensures its essential simplification. This planar HC (Fig. 6.20) contains a base (1) and only two orthogonal carriages (2 and 3), but it also has pairs of limit switches (6 and 7, and 8 and 9) on each controlled coordinate. For each of these coordinates, shift sensors (4 and 5) are connected as before with one of the inputs of operational amplifiers (13 and 14) (Fig. 6.21). As this takes place, each limit switch is connected to the amplifier's second input through an integrator (15 or 16). While performing accurate tasks, the HO uses a position mode by working with carriages whose potentiometers are connected to the operational amplifier's inputs with minimum gain values. If a fast operational mode is needed for the robot, the HO brings the carriages (or carriage) to stress and turns on the limit switches (switch). A linearly building, high-speed signal from the integrator then enters the second amplifier input, which

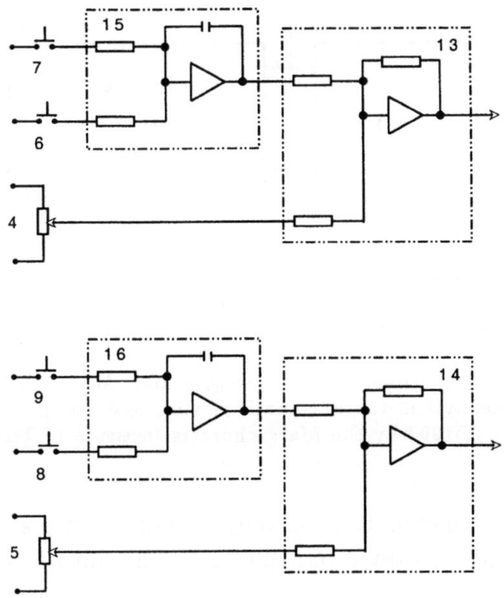

Figure 6.21. Computing device for the HC of Figure 6.20 (6-9 are contacts of limit switches shown in Figure 6.20). From (Slutski, 1997c). ©IMEKO.[1]

is summed with the first signal. Thus, to perform a transportation task, a velocity mode occurs. When the HO takes the carriages from stresses, the signal does not enter the integrator input, and the system returns to a position mode.

This HC computing device (Fig. 6.21) works according to the following algorithm:

$$x_C = k_1 x + (\alpha_1 - \alpha_2) k_2 t, \quad y_C = k_1 y + (\alpha_3 - \alpha_4) k_2 t,$$

where x_C, y_C are the vector components of the robot gripper shift; x, y are the carriage shifts; k_1, k_2 are gains; and $k_1 << k_2$; $\alpha_i, i = 1, ..., 4$ are signals from limit switches (6,...,9) (Fig. 6.20). They are connected as follows: $\alpha_i \in 0, 1,\ \alpha_1 + \alpha_2 = 1,\ \alpha_3 + \alpha_4 = 1$.

By this way, an adaptive control principle is implemented through a simple method. The concept and the purpose of the combined system is realized here with the help of some small changes in the traditional HC. It is obvious that this principle may be used not only for orthogonal carriages of a HC, but also for HCs based on revolute or spherical kinematic pairs, such as joysticks. It is also easy to apply this idea to 3D IDs. Note that such an approach leads to the creation of HCs with small overall dimensions. A

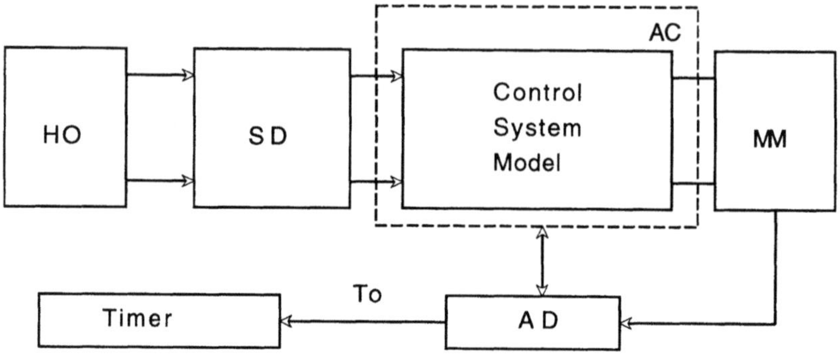

Figure 6.22. Layout of experimental unit. Designations: MM is the manipulator model, SD is the input device, AD is the automatic device, and AC is the analog computer. From (Slutski, 1997b). ©1997 by the Massachusetts Institute of Technology.

peculiarity of these constructions is they do not require an additional input reference element to introduce parameter signals into a control system.

6.4. Simulation and Testing of the Adaptive Systems

Adaptive CPC efficiency testing was realized with the help of semi-natural simulation and the investigation of actual IR control.

6.4.1. SEMI-NATURAL SIMULATION OF ADAPTIVE SYSTEM

The block-diagram of an experimental setup for semi-natural simulation is shown in Figure 6.22. In this setup, there was an HO-controlled, Cartesian, two-coordinate-plane manipulator model (MM) using a plane Cartesian input device (SD) to set control signals that enter the (MM) through an analog computer (AC) in which the control algorithms are realized.

The layout of the control system model is shown in Figure 6.23 [3]. The control system was based on the use of operational amplifiers. Parameter signal control was organized by measuring the squeezing force in handle (H) (consisting of two movable parts). The HO performed a manipulation task; that of moving a model endpoint, in the form of an interchangeable calibrated cylinder, on a trajectory across calibrated holes ($H_1 - H_4$) (Fig. 6.24) drilled into a transparent plane plate. He was required to insert the cylinder into each of the holes in turn. Each hole was provided with an electrical contact ($EC_1 - EC_4$). When the cylinder entered the hole, an automatic device produced an impulse that switched on a lamp, permitting the HO to move on to the next hole. The overall cycle consisted of three shifts (transportation task) and three cylinder guidances to the target (po-

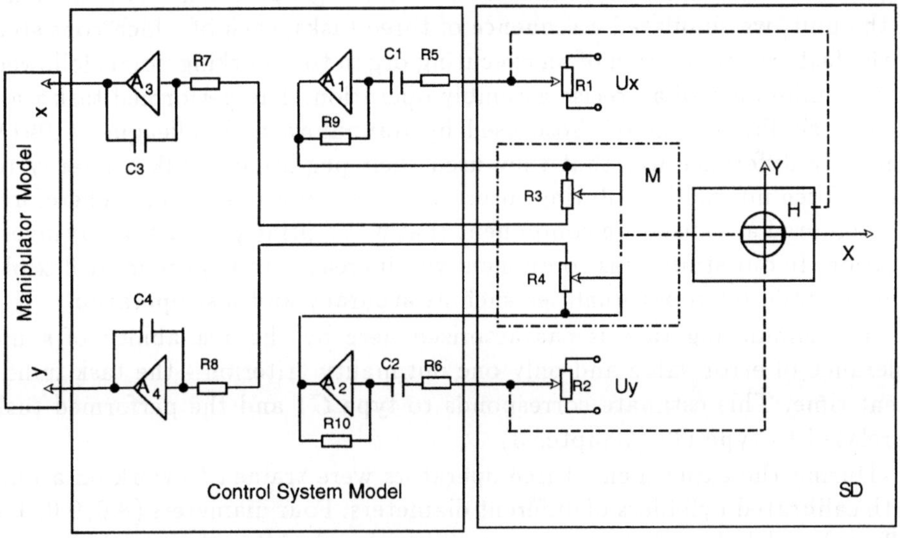

Figure 6.23. Control system layout for experimental investigation of adaptive algorithm. Designations: SD is the input device, H is the control handle, M is the modulator, and $A_1 - A_4$ are operational amplifiers. From (Slutski, 1997a). ©1997 by the American Nuclear Society, La Grande Park, Illinois.

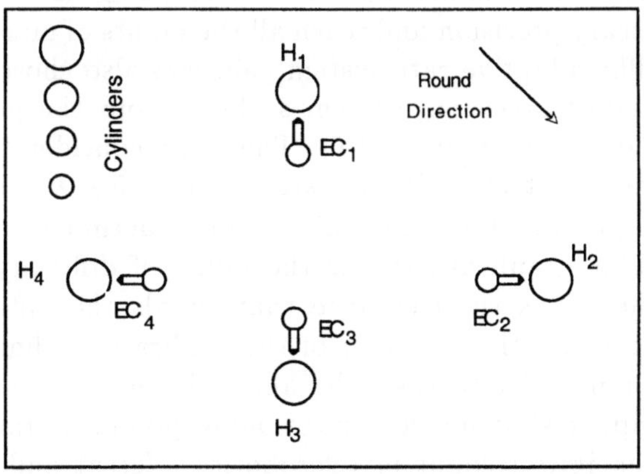

Figure 6.24. Plate with calibrated holes, which serves as the basis of the task fulfilled by the telerobotic system. Designations: $H_1 - H_4$ are calibrated holes, and $EC_1 - EC_4$ are electrical contacts.

sitioning task). The automatic device switched the analog computer on and off, and also the timer that fixed the full-time T_O operation. With the help of the unit, we simulated a sequence of three tasks, each of which consisted of the fast transportation of an executing organ to a working point, followed by the fulfillment of a precise assembly operation. The performed task was, in fact, similar to one of those used by Hannaford and colleagues (1991). The only difference was that they used their peg-in-hole task primarily to study force interactions during insertion/extraction operations. Therefore, they ignored a translation component that was a principal part of our investigation. In our study, this component was increased to require more flexible and versatile telerobot qualities such as accuracy and fast-operation.

The simulating task is characterized here by the availability of some tolerance of error value and only one optimality criterion - the task fulfillment time. This estimate corresponds to type $C1$ and the performed task is related to type $O2$ (Chapter 3).

During the experiment, three operators were trained to work on a unit with calibrated cylinders of different diameters. Four diameters (8.0; 9.0; 9.4; $9.8mm$) and holes of $10mm$ diameter were used. After the training stage, each operator performed 20 tasks for each type of studied control system:

1 - a position coordinate system (6.1);

2 - a position adaptive system (6.7);

3 - a rate adaptive system (6.8).

It is obvious that system characteristics essentially depend on the choice of parameters for each kind of system. Therefore, the coordinate position system gain was set at a minimum $k = 1$, permitting us to simultaneously realize the necessary precision and reach all the points of manipulator model working area. The adaptive rate system gain was also chosen for considerations of necessary precision, system stability, and the possibility of its technical realization ($0.1 \leq k_R \leq 1.0$). The same considerations were used to determine the adaptive position system gain ($0.5 \leq k \leq 5.0$).

Statistically processed average data of the experiment are presented in Figure 6.25. This graph abscissa is the index of difficulty $\log_2(\frac{2S}{\delta})$ (see Chapter 3), where δ is the clearance that equals the difference between the hole diameter and the diameter of the calibrated cylinder inserted by the manipulator model into the hole, and S is the maximum distance of manipulator gripper shift between positioning points. Parameters S and δ are always taken into account in telerobotics (Vertut and Coiffet, 1984), but they are usually considered in isolation. The index of difficulty represents the generalized characteristic that accounts for different kinds of tasks performed by the manipulator.

The curves in Figure 6.25 indicate an increase of control efficiency for adaptive position method. In the investigations, the index of difficulty had

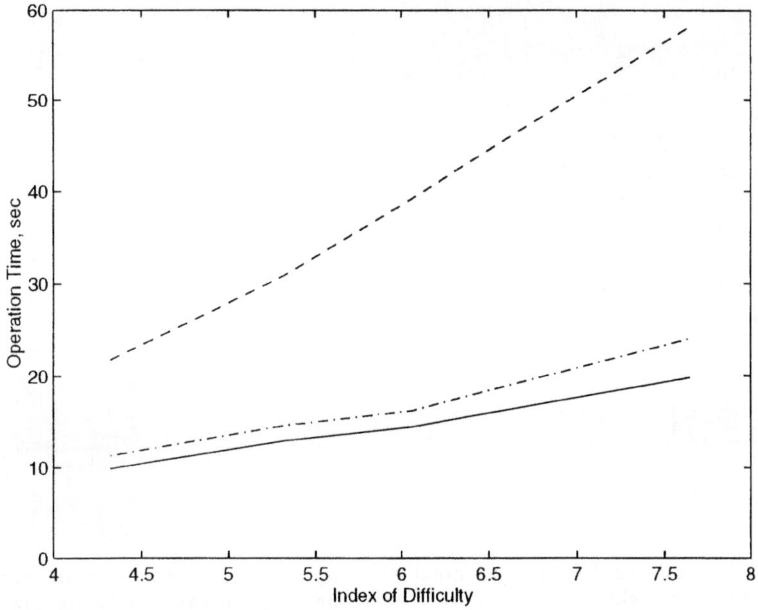

Figure 6.25. Data on the fast-operation of the three kinds of telerobotic systems under fixed accuracy levels. Designations: a solid line - adaptive position system, a dashdot line - coordinate position system, and a dashed line - adaptive rate system.

a maximum value 7.644. For this value, the adaptive system quality index was found to outperform the same index of the coordinate position system by 15%. This indicates that, despite the comparatively small depths of gain control, considerable improvement of operator-manipulator system quality was revealed for the adaptive position system under accuracy allowance decrease.

However, for actual robotic systems, the value S principally coincides with workspace dimensions and the index δ has a value of the same real order. So, for actual robotic systems, the index ν is almost always higher than the above calculated value. Real telerobot operation characteristics are always more to the right of the points showed in Figure 6.25. This indicates an increase of real adaptive control efficiency. Lastly, the dependencies of Figure 6.25 indicate that parameter control does not make the rate system sufficiently effective as compared to the other investigated systems. This leads to a conclusion that the use of such a system for precise task fulfillment is not worthwhile.

An approach utility with automatic gain tuning will be demonstrated separately, while describing results in Chapter 8.

Figure 6.26. Telerobot control terminal with implemented adaptive algorithm and force sensing. From (Slutski, 1997a). ©1997 by the American Nuclear Society, La Grande Park, Illinois.[3]

6.4.2. IMPLEMENTATION AND TESTING OF THE ADAPTIVE CONTROL SYSTEM

The elaboration of remote control systems for actual IR teaching was also performed by the author and his colleagues when he worked in Kazakh State University. One of these systems was implemented (Fig. 6.26) and tested to control a cylinder type IR. This control system was performed in the form of the two separate parts connected by cables: a HC and an electronic control block.

An adaptive position algorithm (6.7) was used in the control block for each remotely controlled coordinate:

$$Y = k_D \int_{t_1}^{t_2} k(t)\dot{X}(t)dt, \qquad (6.20)$$

where $X(t)$ is a coordinate control signal; $k(t)$, a parameter influence; and $k_D = const$, a differentiator gain. Feedback of the forces from the robot EO was also a feature of this control system.

The HC kinematics was chosen as a rocker mechanism in order to obtain a static balance of construction masses (Fig. 6.27). Step-motors (1 and 2) were placed at opposite ends of the rocker (3). The transfer of torque from the step-motor (4) shaft to the handle (5) was achieved with the help of the four-bar linkage ABCD. A gear pair (6 and 7) with the ratio

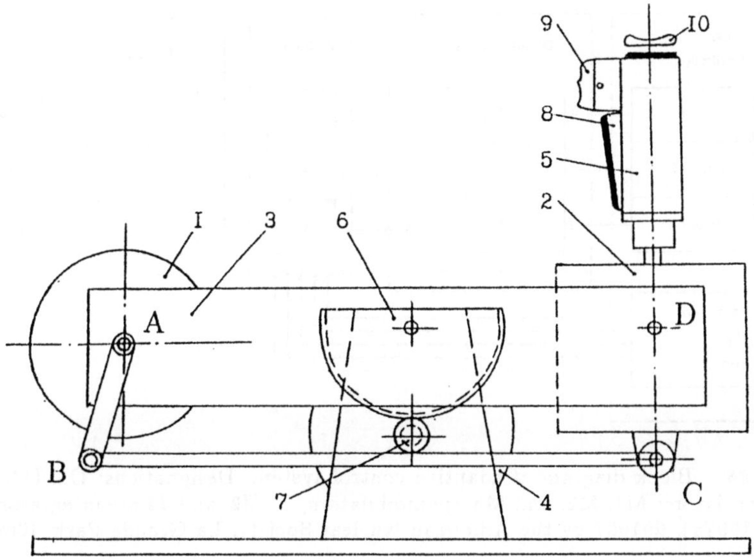

Figure 6.27. HC with static balance of construction masses. Designations: 1, 2, and 4 are step-motors; 3 is a rocker; 5 is a handle; 6 and 7 is a gear pair; 8 is a lever; and 9 and 10 are trigger switches. From (Slutski, 1997a). ©1997 by the American Nuclear Society, La Grande Park, Illinois.[1]

$i = 5.32$ was used to accelerate sensing signal implementation. The handle (5) was attached directly to the step-motor (2) shaft because its inertial moment was small. Position potentiometer sensors were placed on the motor shafts. The handle (5) was provided with a lever (8) of the parameter input reference element. By pressing this lever, the HO switched on five microswitches in turn. Such a layout was described above in Subsection 6.3.2 (Fig. 6.16). Two position trigger switches (9 and 10) were used for on-off control of the robot gripper.

In the electronic converter block, algorithm (6.20) was implemented for three DOFs of the robot's main kinematic chain. The HO was also able to accomplish an on-off control of the robot's orientation DOFs (as gripper pitch and roll) by using special switches placed on the handle.

A force feedback system based on the step-motors is not described here in detail, since this material does not relate directly to the subject of our book.

A block diagram of the electronic control unit is presented in Figure 6.28. This scheme was implemented based on integrated microcircuits. Signals from the three coordinated sensors of the input mechanism enter inputs of three identical control channels.

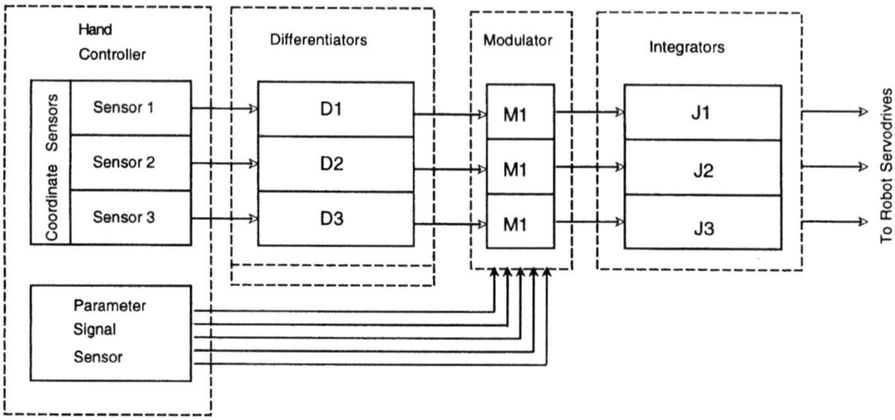

Figure 6.28. Block diagram of adaptive control system. Designations: D1, D2, and D3 are differentiators; M1, M2, and M3 are modulators; J1, J2, and J3 are integrators. From (Slutski, 1997a). ©1997 by the American Nuclear Society, La Grande Park, Illinois.[3]

The input signals enter the differentiators (D1, D2, D3) and then arrive to the modulators (M1, M2, M3). In this block, the parameter signal that arrives from the parameter signal sensor (Fig. 6.28) of the HC is superimposed on the coordinate control signals. The resulting signals are integrated by integrators (I1, I2, I3), and from there go to the power amplifier inputs of the robot servodrives.

During the test, correction depths of the system gains equal to $32db$ were obtained. The test was performed under the control of the robot's (Fig. 6.29) rotation motion when the motion duration from an initial point A to a target point was fixed. This criterion evaluation is related to type $C2$ when it is used for the performance of type $O3$ tasks (Chapter 3). The average indices of positioning accuracy obtained as a result of 10 operations for two regimes were:

- an ordinary coordinate position control (6.1) of 2.3^o, and

- an adaptive position control on algorithm (6.7) of $7.5'$.

Thus, the control accuracy of this manipulation task, using the adaptive system, was more than 15 times greater than with ordinary coordinate control. These results confirm indications gained from semi-natural simulation (see the previous Subsection) that adaptive control efficiency increases when real teleoperations are performed.

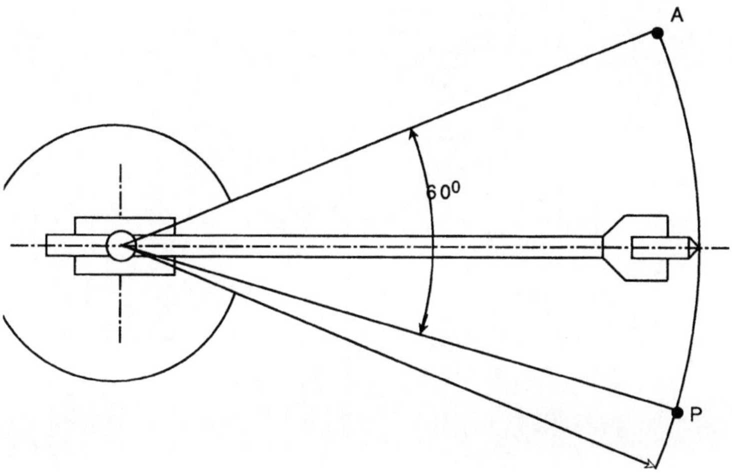

Figure 6.29. IR tested under adaptive and coordinate position control. From (Slutski, 1997b). ©by the Massachusetts Institute of Technology.

6.5. Discussion

The flexible alteration of object parameters provides new possibilities to develop different kinds of effective robots. A meaning of the obtained solution may be explained by Figure 1.3. Whereas the solution based on the use of a coordinate approach allows one to extract a point on this graph that has corresponding values of accuracy and fast-operation, a solution based on the use of an adaptive approach permits one to move a working point along the line while the system is working. Correspondingly, if needed, indices can be flexibly changed within a certain range.

The approach based on the use of the mnemonicability notion for system synthesis permits the organization of adaptive control of a robot's EE position, velocity, and acceleration by using the controlled signals of position, velocity, and acceleration of the input handle. In the latter case, the HO is not limited to a certain place; he/she can move to any point in space. This problem has also been discussed in other studies (Von Hinuber and Janocha, 1993; Weck and Lauffs, 1987) relating to practical instances. A current stage of the elaboration of such an approach may be illustrated by the 6D control device (A Flock of Birds) of the Ascension Technology Corporation (Fig. 6.30), which was used in experiments described in (Nechyba and Xu, 1995). The proposed approach presented in our own work permits this problem to be solved in the context of adaptive control and a sufficiently effective control process of equipment to be organized.

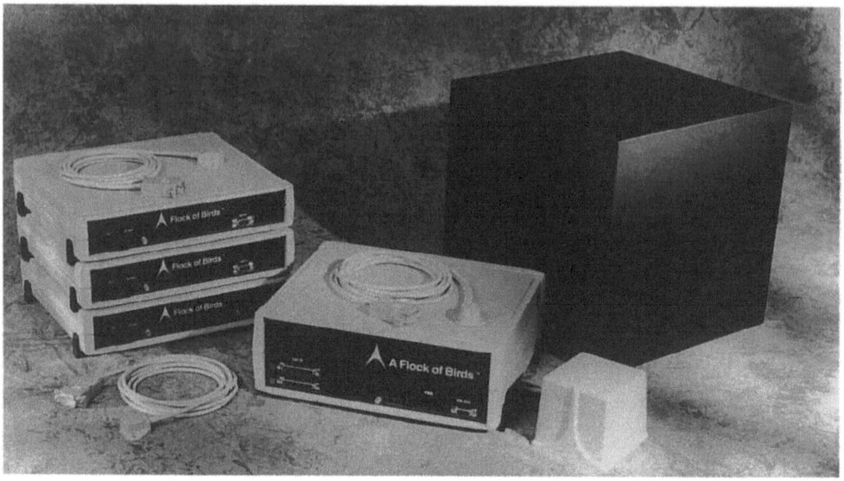

Figure 6.30. "Flock of Birds" components. From left, are three electronics units, one for each of three receivers (sensors). On the right are two transmitter options for the Flock. In the foreground is the standard transmitter, and the larger black box in the background is the extended-range transmitter. In the center of the picture is the controller unit for use with the extended-range transmitter. Courtesy of Ascension Technology Corporation.

All algorithms obtained by us are shown together in Table 6.1, in which conventional coordinate algorithms are also presented for those cases when $k = const$. Conventional algorithms have a traditional form that confirms our assumptions based on the mnemonicability notion which was used above when we obtain adaptive algorithms. Different types of adaptive algorithms may be used depending on conditions and on the purposes of a teleoperation system. For instance, using adaptive algorithms, position-force hybrid systems can be organized.

Adaptive system varieties are proposed that are based either on manual production of the parameter signal, or on automatic production in connection with changes in coordinate control signals. This latter version is less flexible, but it ensures less information loading on the HO. Based on our experience with these systems, we can suggest several fields in which the proposed versions of adaptive control would be useful. As for systems with automatic scaling, they may be effectively used in teaching devices for IRs, because such systems demand the use of simple and reliable IDs, like ordinary joysticks, which also possess small overall dimensions. Systems with manual gain introduction may be basically used for remote control in telerobots because they allow us to organize a more flexible control, whereas rigid requirements for overall dimensions and the complexity of a HC do not always exist in this field.

TABLE 6.1. Algorithms of the adaptive coordinate-parameter control (CPC) and coordinate control (CC) for one controlled channel

Input Signal	Position Control	Control Method Velocity Control	Acceleration Control
CPC ($k = var$)			
Position	$y = \int k(t)\frac{d}{dt}x(t)dt$	$\dot{y} = \int k(t)\frac{d}{dt}x(t)dt$	$\ddot{y} = \int k(t)\frac{d}{dt}x(t)dt$
Velocity	$y = \int k(t)\dot{x}(t)dt$	$\dot{y} = \int k(t)\dot{x}(t)dt$	$\ddot{y} = k(t)\dot{x}(t)$
Acceleration	$y = \int \int [k\ddot{x} + \dot{k}\int \ddot{x}dt]dtdt$	$\dot{y} = \int [k\ddot{x} + \dot{k}\int \ddot{x}dt]dt$	$\ddot{y} = k(t)\ddot{x}(t) + \dot{k}(t)\int \ddot{x}(t)dt$
CC ($k = const$)			
Position	$y = kx$	$\dot{y} = kx$	$\ddot{y} = kx$
Velocity	$y = k\int \dot{x}dt$	$\dot{y} = k\int \dot{x}dt$	$\ddot{y} = k\dot{x}$
Acceleration	$y = k\int \int \ddot{x}dtdt$	$\dot{y} = k\int \ddot{x}dt$	$\ddot{y} = k\ddot{x}$

Remote control with a parameter adaptation permits an increased control system resolution up to the maximum accuracy of position systems; at the same time, it ensures a maximum speed, which is a characteristic of rate systems. Adaptive algorithms permit the use of both analog and digital approaches. Our experience testifies that the first one better ensures real-time required computations. In addition, these algorithms are very simple compared to the combined systems of position-rate control that have the same purpose. As a result, adaptive systems enjoy lower cost and higher reliability.

An important, but perhaps too optimistic, thought is that an application of more precise remote control systems might lower demands for the force reflecting means of telerobots. In fact, while applying the highest precision control with lowest gains in a typical task (e.g., "pig-in-hole"), force interactions may be reduced because one is performing the most precise, finest targeting and orientation, with no contacts between the peg and hole. It is difficult to obtain these lowest gains in conventional semiautomatic systems because of a loss of rate properties of the system; however, adaptive systems permit a flexible combination of precise and rate modes.

Lastly, the distribution of functions between control channels in the adaptive system is as follows:

1) As is normal, the coordinate control circuit ensures realization of the main control goal. For a position type manipulation robot, this means that the desired positioning point is achieved.

2) The parameter control circuit is used to improve (optimize) the con-

trol process quality indices.

It should be noted that the research described here is, in fact, only the first stage of the method development. A number of questions exist to which answers have not yet been found: for example, a stability problem may occur in these systems. In our investigation, this problem was solved experimentally, by choosing a limit of the coefficient $k(t)$ in the algorithm (6.7), because this coefficient value specifically influences system stability. The same approach was used in studying the algorithm with automatic gain adjustment, as described below in Chapter 8. In this case, the parameter α (6.17) was limited in order to remove the danger of instability in the system. However, the proper resolution of this problem waits for further consideration.

Another subject for future investigation is the justification and choice of the most suitable of all possible kinds of adaptive algorithms for use in different industrial applications.

The manipulator flexible scaling may play the same role in remote adaptive control as the impedance adjustment in MSSs. Sheridan (1992) says that this "may be a promising way of making a master-slave teleoperator more versatile." Mention of such an adjustment was also made by Vertut and Coiffet (1984). There is therefore reason to hope that the results presented in this chapter will contribute to the development of ideas for SAC systems put forth by these authors.

References

Hannaford, B., Wood, L., McAffee, D.A., and Zak, H. (1991) Performance evaluation of a six-axis generalized force-reflecting teleoperator, *IEEE Trans. on Systems, Man, and Cybernatics*, **Vol. 21 no. 3**, pp. 620-633.

Horowitz, P. and Hill, W. (1989) *The Art of Electronics*. Cambridge University Press, Cambridge.

Kobrinskii, A.E., Statsenko, V.V., and Tyves, L.I. (1975) Mnemonicability level evaluation of the master-slave manipulator with disconcordance of the input and executing organs, *Mashinovedenie*, **no. 2**, pp. 3-7 (in Russian).

Nechyba, M.C. and Xu, Y. (1995) Human-robot cooperation in space: SM^2 for new space station structure, *IEEE Robotics and Automation Magazine*, **Dec.**, pp. 4-11.

Sheridan, T.B. (1992) *Telerobotics, Automation, and Human Supervisory Control*. The MIT Press, Cambridge, MA.

Slutski, L.I. (1979) Operative correction at manipulator velocity control, *Proc. of 1st Interinstitute Conf. Robotics. Control and Sensitive Systems*, **Vol. 1**, Kaunas (USSR), pp. 36-40 (in Russian).

Slutski, L. (1994) Telerobotic systems with manipulator parameter control, *Proc. of Fourth Intern. Conf. on Artificial Reality and Tele-Existence*, July 14-15, Tokyo, pp. 237-253.

Slutski, L. (1995) Telerobot adaptive control, *Proc. of 3rd French-Israeli Symp. on Robotics*, Herzeliah, pp.92-96.

Slutski, L. (1996) Adaptive telecontrol for industrial robot teaching, *Proc. of 27th Intern. Symp. on Industrial Robots*, Milan, Italy, 6-8 October, pp. 47-52.

Slutski, L.I. (1997a) Adaptive control techniques for teleoperated manipulators, *Proc. ANS 7th Topical Meeting on Robotics and Remote Systems*, Augusta, Georgia, pp. 661-667.

Slutski, L. (1997b) On-line telecontrol techniques based on object parameter adjustment, *Presence*, **Vol. 6 no. 3**, pp. 255-267.

Slutski, L. (1997c) Control means of adaptive teleoperators. *Proc. of XIV World Congress*, **Vol. IXB**, 17, Topical Workshop on Virtual Reality and Advanced Man-Machine Interfaces, Tampere, Finland, pp. 1-7.

Slutski, L. (In press) Design principles of adaptive control systems for telerobots, *Robotics and Computer-Integrated Manufacturing.*

Slutski, L. and Coiffet, P. (1996) Man-adaptive control of teleoperators, *Robotica*, **Vol. 14**, pp. 111-117.

Vertut, J. and Coiffet, P. (1984) *Teleoperation and Robotics: Evolution and Development*, Vol.3A. Prentice-Hall, Englewood Cliffs, N.J.

Von Hinuber, E. and Janocha, H. (1993) Inertial measurement system for calibration and control of robots, *Industrial Robot*, **Vol. 20 no. 2**, pp. 20-27.

Weck, M. and Lauffs, H. (1987) Sensoren beschleunigen on-line-roboterprogrammierung, *Industrie-Anzeiger*, **Vol. 109 no. 8**, pp. 19-22 (in German).

Notes

[1] The figure appeared also in (Slutski, to be appear); ©Elsevier Science Ltd, The Boulevard, Langford Lane, Kidlington 0X5 1GB, UK.

[2] The first figure appeared also in (Slutski, to be appear); ©Elsevier Science Ltd, The Boulevard, Langford Lane, Kidlington 0X5 1GB, UK.

[3] The figure appeared also in (Slutski, 1997b); ©1997 by the Massachusetts Institute of Technology.

CHAPTER 7

MECHANICAL MEANS FOR REALIZATION OF ADAPTIVE CONTROL

Manipulation Devices with Variable Link Lengths

7.1. Basis of Approach

In some cases, another approach to adaptive CPC making is required due to the impossibility to implement the described principle of adjustable scaling in a manipulator control system. For example, in MSSs, it leads to deterioration of the mnemonicability and, as a consequence, of system accuracy. Another method of CPC implementation may then be proposed based on the fact that the resultant system gain is the product of the gain of the system control part (k_c) and the gain of its mechanical part (k_m)

$$k = k_c \cdot k_m.$$

Thus, if the control of the value k_c is impossible, it may still be possible to control the parameter k_m. Practically, the latter is the manipulator scale ratio. In MSSs, this ratio may, in turn, be controlled by one of two methods. The first utilizes scale alteration of the slave arm mechanism.

7.2. Slave with Variable Link Lengths

One possibility of implementing this control approach is related to carrying out slave main links with variable lengths. An example of such a manipulator is shown in Figure 7.1. Its ordinary anthropomorphic structure feature is the availability of the shoulder and forearm with variable lengths. Certain drive systems for these length changes should be used in this manipulator. These will be discussed below, together with two possible operational regimes of the manipulator.

7.2.1. MODE I - SYNCHRONIC CHANGE OF LINK LENGTHS

The first regime is based on synchronic changes of manipulator link lengths. Let us to explain this regime by using Figure 7.2,[1] which represents an example of the master-slave control of a planar manipulator. The master and the slave of this manipulator are geometrically similar to each other and

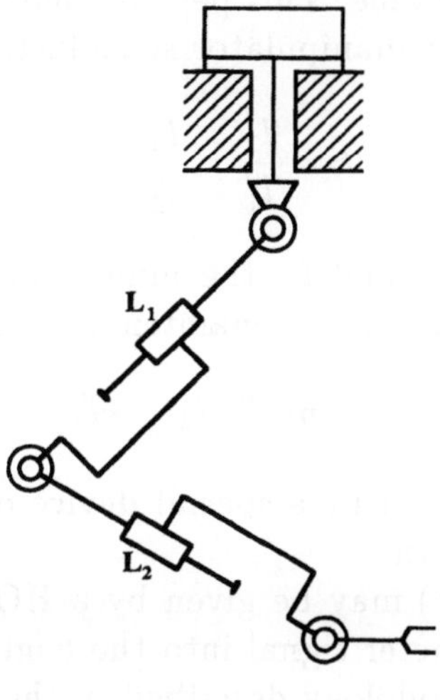

Figure 7.1. Slave with variable link lengths L_1 and L_2.

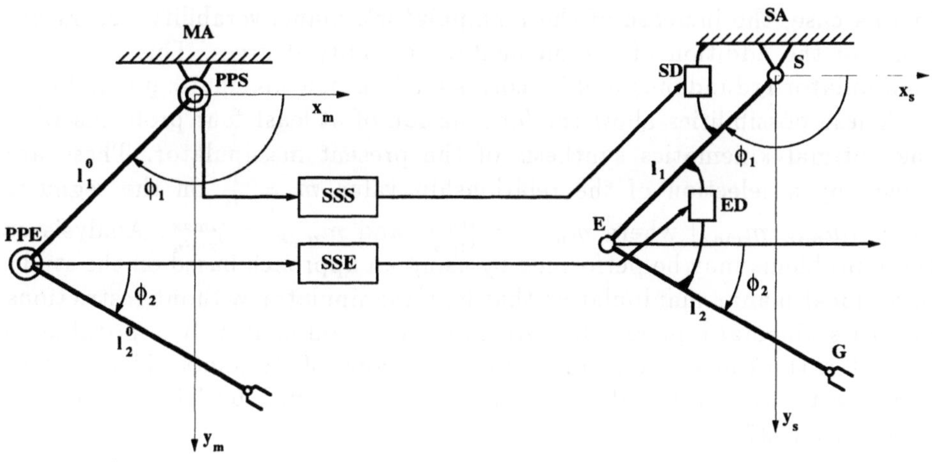

Figure 7.2. Example of a MSS with variable link lengths of the slave. Designations: MA is a master arm, SA is a slave arm, PPS and PPE are position sensors, SD is a shoulder drive, ED is a elbow drive, and SSS and SSE are servosystems. From (Djoldasbekov and Slutski, 1983).[1]

contain two degrees of mobility in the shoulder (S) and elbow (E) joints. Master links are made with constant lengths l_1^0 and l_2^0, which are connected with each other by a constant relationship $\frac{l_1^0}{l_2^0} = m_0$. Slave link lengths are connected by a similar relationship $\frac{l_1}{l_2} = m$ and, initially, $m = m_0$; however these lengths are made variable. This permits one to change the lengths to obtain various values of the manipulator scale factor (see Subsection 4.4.2)

$$k = \frac{l_1}{l_1^0} = \frac{l_2}{l_2^0}. \tag{7.1}$$

With the length changes l_1 and l_2, the mnemonic master-slave control is preserved because of the automatic maintenance of the relationship

$$l_1 = km_0 l_2^0,\ l_2 = kl_2^0. \tag{7.2}$$

These conditions are satisfied by a special device of program parametrical control of the SA link lengths.

The scale factor k (7.1) may be given by a HO by the introduction of the complementary parameter signal into the control system according to the adaptive control methodology described in the previous Chapter.

7.2.2. MODE II - INDEPENDENT CHANGE OF THE LINK LENGTHS

In this case, the increase of the manipulator's maneuverability occurs because of the addition of complementary mobility degrees. This results in manipulator redundancy and its corresponding new kinematic possibilities.

These possibilities allow the formulation of at least four problems with the optimal kinematics synthesis of the present manipulator. These are solved by a selection of the relationship value $m = \frac{l_1}{l_2}$ on the segment $M = [m_{min}, m_{max}]$ where $m_{min} = \frac{l_{1min}}{l_{2max}}$ and $m_{max} = \frac{l_{1max}}{l_{2min}}$. Analysis of these problems may be performed by using an approach based on the study of an ideal planar manipulator; that is, a manipulator without restrictions in all its kinematic pairs. Moreover, for this manipulator, it is possible to neglect a thickness in its links and dimensions of its joints. This part of our investigation is based on the approach developed by Vinogradov and colleagues (1971).

Problem 1

To provide the maximum value of the volume V of this manipulator's workspace, when selecting the value $m = m_1$

$$\sup_{m \in M} V = V(m_1). \tag{7.3}$$

Let us to assume, for simplicity, that the initial relationship of link lengths for the manipulator is $l_1 > l_2 > l_3$, which is maintained constantly during manipulator work. The maximum distance of the gripper (G) from the shoulder (S) is then (Fig. 7.3) $R_{max} = l_1 + l_2 + l_3$, and the minimum,

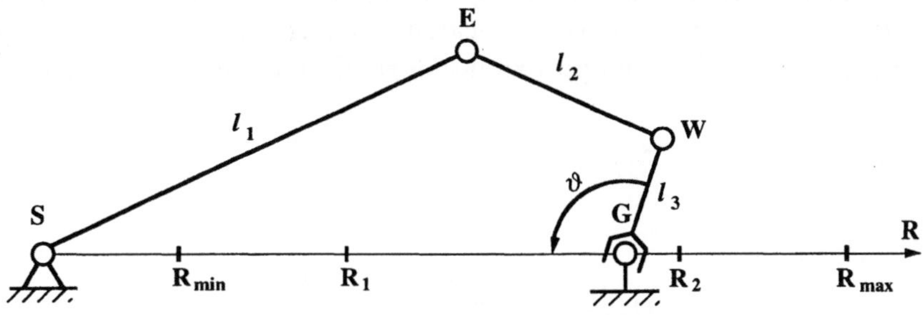

Figure 7.3. Kinematics of an ideal planar manipulator (S is a shoulder, E is a elbow, W is a wrist, and G is a gripper). From (Djoldasbekov and Slutski, 1983).[1]

$R_{min} = l_1 - l_2 - l_3$. A dextrous workspace (Vijaykumar *et al.*, 1985) is situated then between values R_1 and R_2. The manipulator servicing area (maximal workspace) is then

$$S = \pi(R_{max}^2 - R_{min}^2) = 4\pi l_1(l_2 + l_3). \tag{7.4}$$

It is obvious that $\sup_{l_1,l_2} S$ is obtained when $l_3 = const$ and the values $l_1 = l_{1max}$ and $l_2 = l_{2max}$; i.e., $m_1 = \frac{l_{1max}}{l_{2max}}$.

Problem 2
To find the parameter value $m = m_2$ which enables us to obtain the maximum value of the service factor Ω (which characterizes manipulator dexterity (Vijaykumar *et al.*, 1985)) of a certain point in the manipulator's workspace

$$\sup_{m \in M} \Omega = \Omega(m_2). \tag{7.5}$$

According to the definition of Vinogradov et al. (1971), the service factor of a certain point of a workspace is equal a ratio $\Omega = \frac{\psi}{4\pi}$ where ψ is a certain solid angle inside of which a manipulator EE may reach the point.

For the case of a planar manipulator (Fig. 7.3), the service factor is the following function of a service angle ϑ

$$\Omega = \frac{1 - \cos\vartheta}{2},$$

where ϑ is a maximum angle of the link l_3 rotation about point G for given lengths of arm mechanism.

In the case of the ideal planar manipulator, we may apparently speak about the maximization of the factor Ω (7.5) on the zones of incomplete service of the manipulator's workspace. As is known, there are two such zones (Fig. 7.3). Zone I is located by the value $R = R_{min}$ up to the value $R = R_1$. The other, zone II, is located from the value $R = R_2$ up to $R = R_{max}$. In zone II, the approximate formula for the service factor is:

$$\Omega_{II} = \frac{(l_1 + l_2)^2 - (R - l_3)^2}{4Rl_3}, \tag{7.6}$$

and in zone I,

$$\Omega_I = \frac{(R + l_3)^2 - (l_1 - l_2)^2}{4Rl_3}. \tag{7.7}$$

It is evident that Ω_{II} is realized at maximum lengths values l_1 and l_2; i.e.,

$$m_2^{II} = \frac{l_{1max}}{l_{2max}}.$$

In zone I (7.7), the condition for the best service of workspace points is the realization of the relationship $l_1 = l_2$; i.e., the value $m_2^I = 1$.

Problem 3

To maximize the average service factor in manipulator workspace by changing the value

$$\sup_{m \in M} \overline{\Omega} = \overline{\Omega}(m_3). \tag{7.8}$$

For an ideal planar manipulator, the average service factor may be written as follows:

$$\overline{\Omega} = \frac{1}{s} \int_{(S)} \Omega dS.$$

Here, S is the working manipulator's area and Ω the service angle of a certain point in the working area.

A change of the argument R allows us, when using (7.6) and (7.7), to conclude the simplified variant of the corresponding formula, which finally looks like

$$\overline{\Omega} = 1 - \frac{l_3}{l_2 + l_3}.$$

It is evident now that the supremum of $\overline{\Omega}$ is obtained at the value $l_2 = l_{2max}$, and that the initial relationship $l_1 > l_2$ gives the second condition in the form of $l_1 > l_{2max}$. Thus, for the studied problem the value of m_3 parameter lies in the following range

$$1 < m_3 \leq \frac{l_{1max}}{l_{2max}}.$$

Problem 4
To find the value m that minimizes the volume V_{st} of the "dead" zones of the manipulator (stray volume):

$$\inf_{m \in M} V_{st} = V_{st}(m_4). \tag{7.9}$$

By stray volume, we understand the difference between the volume V_w of manipulator workspace and V_s- servicing volume in which the manipulator operates

$$V_{st} = V_w - V_s.$$

The stray area of the planar ideal manipulator, as seen in Figure 7.3, is a circle with the radius $R = R_{min}$. Therefore, obtaining the infimum of this radius corresponds to the minimization of the value S_{st}. The infimum is provided by the arbitrary value of one of the changeable links - for instance, $l_2 = \tilde{l}_2$ - by selecting a corresponding length of an other link using the relationship

$$\tilde{l}_1 = \tilde{l}_2 + l_3.$$

The desired parameter m then has the value

$$m_4 = 1 + \frac{l_3}{\tilde{l}_2} < 2,$$

because $l_2 > l_3$.

7.2.3. CONSTRUCTIVE REALIZATION

Concerning development of a drive system to change manipulator link lengths (Fig. 7.1), a straightforward solution is associated with placing special drives for each additional mechanism (L1 and L2). However, another approach is possible that allows the simplification of manipulator construction by using simpler commutative elements instead of special drives.

The explanation of this idea which follows is based on some of our results.

One of the possible solutions makes use of the joint-lever mechanisms of pantographs (Figs. 7.4 and 7.5) as the arm's links (shoulder and forearm). Figure 7.4 shows the manipulator arm in the base plane (the mechanism of the arm's rotation in the shoulder joint about the vertical axes is not presented). The pantographs' joint-bar outline with variable configuration allows the regulation of the length of each link of the arm. Motion translation to the pantographs is brought about by the special motor (11) via the gear transmission with the drive gear (13) and driven gears (16) and (17).

Implementation of the required modes of the manipulator's operation is obtained using commutation elements - friction electromagnetic clutches

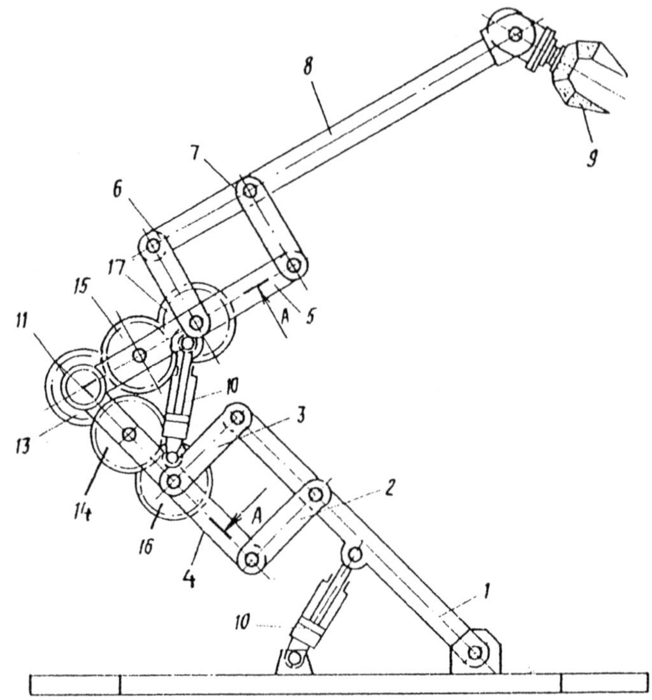

Figure 7.4. Manipulator with variable lengths of links in the form of pantographs. The pantographs of shoulder and forearm consist of details (1-4) and (5-8), respectively. The manipulator contains also a gripper (9), driving motors (10), a special driving motor (11), and gears (13-17). From (Djoldasbekov and Slutski, 1983).[1]

(Fig. 7.5). The variation of link lengths in this construction is implemented by the driving motor (11). For this, the gear (16) is connected to link (3) via electromagnetic clutch (18), and the gear (17) is connected to link (6) via electromagnetic clutch (19). Clutches (20) and (21) serve to fix relative positions of the links (3 and 4) and (5 and 6).

The manipulator works as follows:

The motor (11) turns intermediate gears (14) and (15) through clutch (12) of a drive shaft and a gear (13). Gears (14) and (15) pass rotation to gears (16) and (17) which are rigidly connected to driven shafts. A further manipulator work depends on putting on modes of clutches (18, 19, 20), and (21). The EO may work in four modes depending on combinations of the clutches being put on.

The first mode

If clutches (20) and (21) are on, and clutches (18) and (19) are off, rotation moments of clutches (16) and (17) are not passed to links (3)

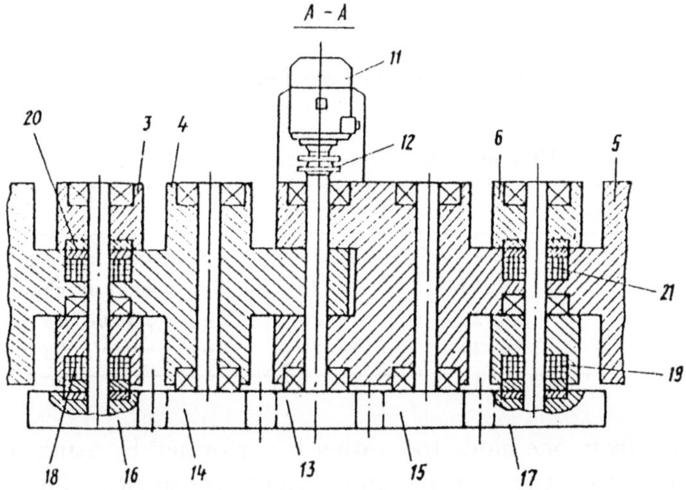

Figure 7.5. A section A-A through the manipulator arm of Figure 7.4. The designations are the same as those of Figure 7.4 and, in addition, 18-20 are friction electromagnetic clutches.

and (6). Link (4) is rigidly connected to link (3), and link (5) is rigidly connected to link (6). The manipulator then works at the conventional mode for manipulators with constant link lengths.

The second mode

If clutches (18) and (19) are on, and clutches (20) and (21) are off, rotation moments are passed from gears (16) and (17) to links (3) and (6). As this takes place, shoulder and forearm lengths are synchronously changed.

The third mode

If clutches (18) and (21) are on, and clutches (20) and (19) are off, the rotation moment is passed from gears (16) to link (3); but from gear (17), it is not passed to link (6). Link (5) is rigidly connected to link (6). As this takes place, the shoulder length is changed, and the forearm length is not changed.

The fourth mode

If clutches (18) and (21) are off, the rotation moment is then passed from gears (17) to link (6); but from gear (16), it is not passed to link (3). Link (4) is rigidly connected to link (3). As this takes place, the forearm length is changed, and the shoulder length is not changed.

A feature of this construction is, while working in modes 3 and 4, changes of shoulder and forearm lengths are independent relative to each other.

Data about these modes are shown together in Table 7.1.

TABLE 7.1. Switchings of the clutches and work modes for the manipulator of Figures 7.4 and 7.5. Designations: L_S is the shoulder length and L_F is the forearm length.

Clutches Regimes	Clutches On	Clutches Off	Comments
Kind of the Mode			
Mode I	20, 21	18, 19	Constant lengths of links
Mode II	18, 19	20, 21	Synchronously changed links
Mode III	18, 21	19, 20	$L_S = variable$, $L_F = constant$
Mode IV	19, 20	18, 21	$L_F = variable$, $L_S = constant$

A passage from one mode to another is performed by using corresponding commutations in the control system, which lead to switching the clutches. However, the passages should be performed in agreement with link lengths required.

It should be noted that application of the special motor (11) is not necessary. Link lengths variability may be realized by means of the same drive motors (10) installed in the EO joints. This idea will be developed below.

A further development of this approach is associated with a performance of the parts, that form a movable contour pantograph, whose links can also be varied in length. Such a scheme is presented in Figure 7.6. Its principal difference from the manipulator shown in Figure 7.4 is in the parts of the pantograph contour - the hydraulic cylinders (4-7) and (10-13). This increases the possibility to improve robot kinematic properties because of the added ability to increase robot link lengths. This is not only because of changes in pantograph configuration, but also because of a change in the lengths of the parts of the pantograph. In this case, there is a change of the lengths of the manipulator links, as well as of the links' shapes. This increases also robot maneuverability and carrying capacity.

However, this approach, which is based on the use of pantographs as robot arm links, leads, perhaps, to a too-complicated construction of a robot arm. Therefore, another solution of the problem is possible - one that is based on the use of telescopic manipulator links. A sample of this kind of manipulator is shown in Figure 7.7. This variation contains a drive motor (5) with a reducer and telescopic links that consisted of parts (1, 2) and (3, 4). The motor (5) is connected to a shaft (6) with bevel gears (7 and 8). Gears (9-12) are situated on the shaft (6) through plain bearing. The device also contains bevel gears (13 and 14), which together with the gears (7 and 9) and (8 and 12), form a pair of planetary gear trains.

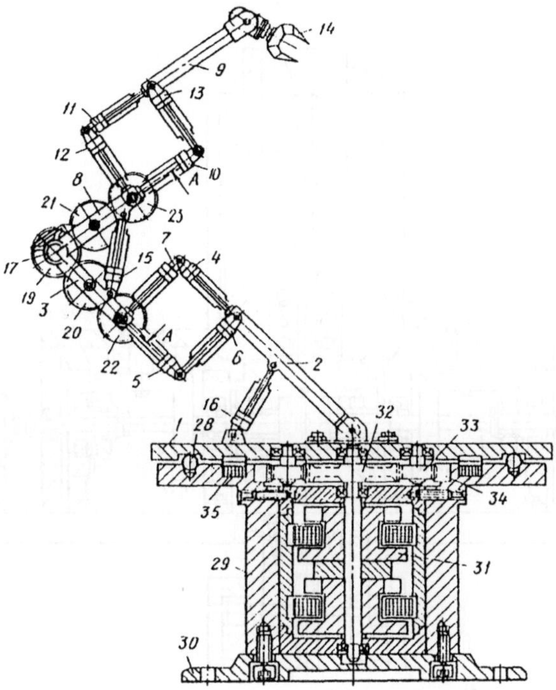

Figure 7.6. Manipulator with variable lengths of pantographic links (4-7 and 10-13 are hydraulic linear power motors).

Gears (10 and 11) are engaged to a bevel gear (15) which is rigidly bound to a worm (16) of the worm-gear-drive, whose drive nut (17) is rigidly bound to part (1) of one of the manipulator links. Gear (9 or 12) connects gear (10 or 11) to the shaft (6) with the help of the frictional electromagnetic clutches (18 or 19) and (20 or 21), respectively. Fixation element (22) joins the parts (2 and 3) of the EO links, if needed.

The motor (5) of the EO turns the bevel gears (7) and (8) via the shaft (6). The gears pass rotation to gears (9 and 12) through bevel gears (13 and 14). Further work of a joint drive depends on modes of clutches (18-21) adopted.

The first mode

If clutches (18 and 19) are on, clutches (20) and (21) are off, the gears (9) and (12) are stopped, because they are rigidly bound to the part (3). The rotation moments of gears (10 and 11) do not pass to the bevel gear (15). The drive works according to a basic regime with constant link lengths. Direction changes of link rotation in the joint is obtained by reversing the motor (5).

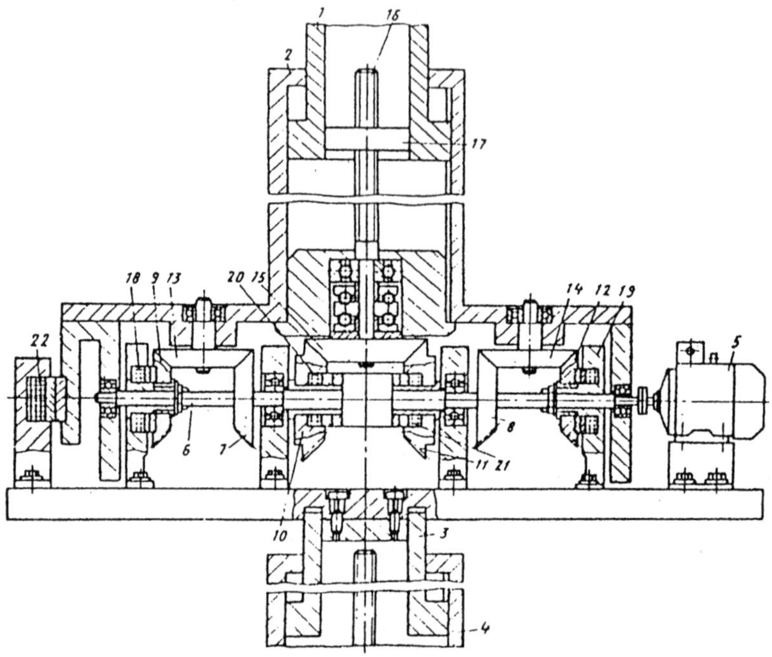

Figure 7.7. Manipulator with variable lengths of telescopic links. The principal designations are: 1, 2 and 3, 4 are telescopic links; 5 is a drive motor; 7-15 are gears; 16 is a worm; and 18-21 are frictional electromagnetic clutches.

The second mode

If clutch (20) (or (21)) and element (22) are on, clutches (18 and 19) and (21) (or (20)) are off, the rotation moment is passed from gear (10) (or (11)) to worm (16) of the worm-gear-drive through bevel gear (15). A length of the link, which consists of parts (1) and (2), is then changed, but the relative position of the manipulator links remains constant.

The third mode

The third mode consists of a combination of two first modes. If clutches (18 and 19) and (20) (or (21)) are then on, clutch (21) (or (20)) is off, this ensures simultaneous rotation of the links in the manipulator joint and a length change of the link, which consists of parts (1 and 2). While reversing the motor (5), a rotation direction of the manipulator links is changed; a change in the switching of clutch (20) (or (21)) also occurs in this case.

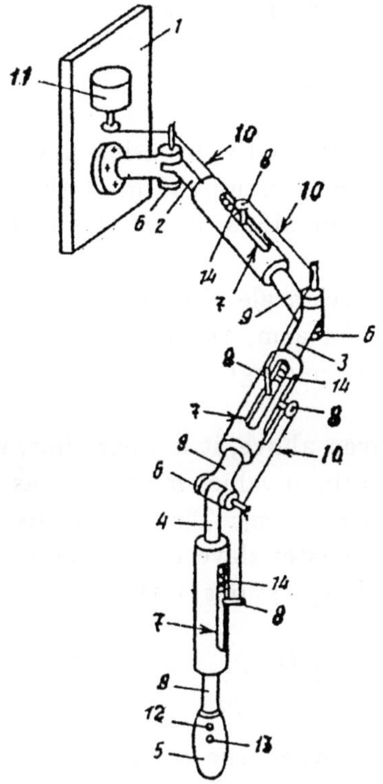

Figure 7.8. Master with variable lengths of telescopic links. It contains a base (1), joint-connected telescopic links (2-4), and a handle (5). Position sensors (6) are situated in the joints. Slots (7) are cut in links (2-4), and the fingers (8) are placed there and firmly attached to the inner elements (9) of the telescopic links. Cables (10) connect the fingers to the motor (11). From (Slutski and Coiffet, 1996). ©Cambridge University Press.

7.3. Master with Variable Link Lengths

The problem can also be solved by having variable lengths of the master links (Slutski and Djamalov, 1989). Such a master layout is shown in Figure 7.8. The motor (11) is specifically used here to modify lengths of the master links.

Controlling this master, the HO moves its links (2-4) as usual by using the handle, and the signals to the slave control are taken from the joint sensors. When it is necessary to use the slave at high velocity, the HO increases the manipulator scale ratio by pressing button (12). A movement from motor (11) is passed to the inner elements of the links, and these elements are pushed into the links, which are proportionately shortened. If it is necessary to carry out a high-accuracy operation, the HO presses button (13) for the reverse motion of the motor. The springs (14) then pull out elements (9), the link lengths increase, and the scale ratio decreases.

7.4. Discussion

The way to organize adaptive CPC by using mechanical means described above leads to the creation of manipulators with variable scale ratio. This approach is only suitable for MSSs, where this ratio may be controlled by one of two methods.

The first method utilizes scale alteration of the slave arm mechanism. In order to implement the arm, the lengths of its links must be made variable; for example, with pantographic or telescopic shape (Djoldasbekov and Slutski, 1983).

From the analysis given above, it is clear that, when solving Problems 1-4 (Subsection 7.2.2), optimal values m_i do not, as a rule, equal the initial value of the relationship $m = m_0$. This fact leads to the violation of the mnemonicability of the master-slave control (Fig. 7.2), which is evident from the consideration of expressions of the system gains at two coordinate axis:

$$k_x = k\frac{m\cos\phi_1 + \cos\phi_2}{m_0\cos\phi_1 + \cos\phi_2}; \quad k_y = k\frac{m\sin\phi_1 + \sin\phi_2}{m_0\sin\phi_1 + \sin\phi_2}.$$

Hence, we see that, only when the condition $m = m_0$ is satisfied does the equality $k_x = k_y = k$ occur. In the opposite case, gains k_x and k_y become different functions of the generalized coordinates ϕ_1 and ϕ_2, which leads to the violation of control mnemonicability. A certain loss of the operator-manipulator system quality then occurs (Chapter 3). This loss is a kind of compensation for obtaining optimal kinematic properties for the manipulator. However, the level of the work quality can easily be restored by returning to the initial manipulator's kinematics characterized by the value $m = m_0$.

Under co-ordinate motion of the drives for the link lengths alteration, a manipulation system scale ratio is changed and, accordingly, its quality indices change. It is obvious that, in this case, the slave construction gets rather complicated; but achievable effect is not only limited by the advantages of the CPC approach. A manipulator is also able to perform a number of new technological operations, because of kinematic redundancy in the arm structure. Variable link lengths in the constructions described in this Chapter make the manipulator EO more versatile, which provides the possibility of its multi-objective utilization. The use of a slave with variable link lengths enables, within the range of one constructive realization, the transition from one geometry of the EO to another. This can result of a wide range of manipulation operations which are optimally produced by a teleoperator. Such manipulators also allow us to avoid collisions and to organize effective motions of a robot arm in environments with obstacles (Kolarov and Roth, 1992).

Another approach to solve the problem described above is to make the links of the master arm of variable lengths (Section 7.3). A change in the manipulation system quality indices also occurs, but this method is indeed simpler and cheaper than the previously described implementation of the system with variable link lengths of the slave arm. Although the kinematic advantages of the first solution (Section 7.2) are lost, the simplicity of the system may often be more important.

From a control standpoint, an important overall feature of the approach presented in this Chapter is that the coordinate and parameter control channels are operated separately; that is, the second channel acts independently from the first (because they control different drives). This differs from the approach of the previous Chapter, where the two channels interact with one another.

References

Djoldasbekov, U.A. and Slutski, L.I. (1983) Manipulators with variable link lengths: Kinematics and possibilities, *Mechanism and Machine Theory*, **Vol. 18 no. 4**, pp. 271-274.

Kolarov, K. and Roth, B. (1992) Best estimates for the construction of robots in environments with obstacles, *Proc. IEEE Intern. Conf. on Robotics and Automation*, Nice, France, pp. 377-382.

Slutski, L. and Coiffet, P. (1996) Man-adaptive control of teleoperators, *Robotica*, **Vol. 14**, pp. 111-117.

Slutski, L.I. and Djamalov, N.K. (1989) Master arm of a master-slave manipulator, USSR author's certificate # 1484677, *Bulletin of Inventions*, **no. 21** (in Russian).

Vijaykumar, R., Tsai, M.J. and Waldron, K.J. (1985) Geometric optimization of manipulator structures for working volume and dexterity, *Proc. IEEE Intern. Conf. on Robotics and Automation*, St. Louis, pp. 228-236.

Vinogradov, I.B., Kobrinskii, A.E., Stepanenko Y.A. and Tyves, L.I. (1971) Peculiarities of manipulator kinematics and method of volumes, *Mechanica Mashin*, **Vol. 27-28**, pp. 5-16 (in Russian).

Notes

[1] Reprinted here with kind permission from Elsevier Science Ltd, The Boulevard, Langford Lane, Kidlington 0X5 1GB, UK.

CHAPTER 8

METHODOLOGIES FOR ANALYSIS AND DESIGN OF TELECONTROL SYSTEMS

Approaches to Analysis of Remote Control Processes

8.1. Introduction

In Chapter 3, it was mentioned that an important problem concerning telecontrol is the need to conduct many experimental investigations when designing new equipment. This increases elaboration time and cost. Therefore, development of new computer-based design methods for MMSs (and, in particular, telerobotic systems) is very important.

Among such methods the most promising are those based on Fitts's law (1954). However, these results do not always permit sufficient evaluation of system quality; such evaluations are not precise and are based on a very general description of HO motion characteristics. Furthermore, resultant criterion dependences do not always correspond to the design problem of a concrete remote control system, and their application is usually too complicated. In this connection, the need to find answers to such problems places in the forefront the necessity for profound studies of HO behavior. The pursuit of such studies may lead to more precise descriptions of system quality characteristics, as well as to the design of promising and effective MMSs.

The present Chapter proposes some new methodologies to evaluate and predict telerobot quality with the help of some models of human behavior. The first approach was applied to the study of remotely controlled machines with typical position and rate algorithms (Chapter 4). The second method was used to investigate an adaptive algorithm (Chapter 6) for telerobot operations.

8.2. Method Based on Adaptive HO Model

8.2.1. BASIS OF APPROACH

As stated earlier (Subsection 3.3.2), the most challenging problem when modeling manually controlled machines is the absence of effective HO models that account for human adaptive possibilities relative to the dynamics of a controlled machine (robot). In this connection, an adaptive HO model was proposed by us (Kobrinskii *et al.*, 1971), and further developed (Djoldasbekov *et al.*, 1986), which provides for a united approach to MMS calculation and design.

This model is based on two assumptions:

- A) for a given machine, the HO elaborates an optimal control law according to a complex criterion (3.4). For instance, while performing a targeting task using precision guidance (the error $\Phi = const$ and a function of the task completion time $F(T) = T \rightarrow min$) (see Chapter 3), he/she elaborates a time-optimal control law;
- B) the HO has physiological limitations: a pure delay δ_1, an error in time of control law switchings δ_2, and limitations on all kinds of human output motion derivatives

$$|\dot{x}| \leq M, \; |\ddot{x}| \leq R, \; |\frac{d}{dt}\ddot{x}| \leq L, \ldots \tag{8.1}$$

  which prevent realization of the optimal control law. These limitations result in distortion of positioning errors and changes in task fulfillment time. Note that limitations M, R, and L are, in fact, statistically determined mean values of random variables.

The assumption of a time-optimal human-control model was confirmed for simpler kinds of position- and rate-type machines in our studies (Kobrinskii *et al.*, 1971; Kobrinskii *et al.*, 1974). For more complicated machines, the same was demonstrated by Bulgarian scientists (Gidikov *et al.*, 1973).

Thus, a principal part of the methods used with the above HO model to obtain characteristics of MMS accuracy and fast-operation includes the following steps:

- determination of a time-optimal control law for this type of machine (by introducing an adaptive property into a model, relative to controlled-object dynamics);
- deformation of this control law because of the limitations noted in (8.1);
- determination of the MMS output signal that corresponds to the obtained input-deformed signal;
- determination of the output error and task-fulfillment time for the machine endpoint.

This approach was used to determine quality indices for previously described types of inertialess, one-coordinate machines:

a position-type machine

$$y = kx, \; k = const \tag{8.2}$$

and a rate-type machine

$$\dot{y} = \eta x, \; \eta = const, \; 0 \leq x \leq N^*, \tag{8.3}$$

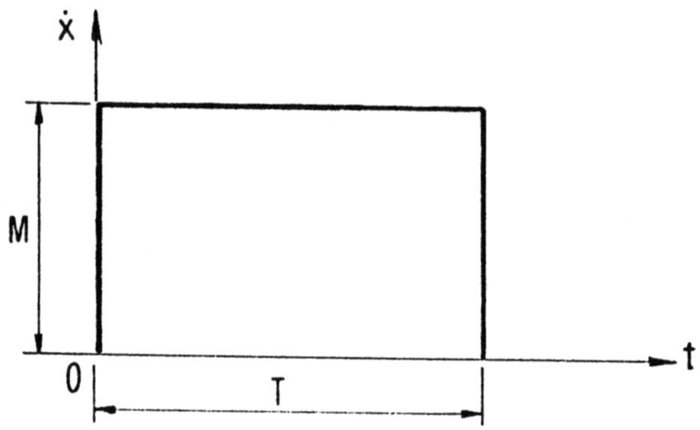

Figure 8.1. Optimal input control signal for position-type machine.

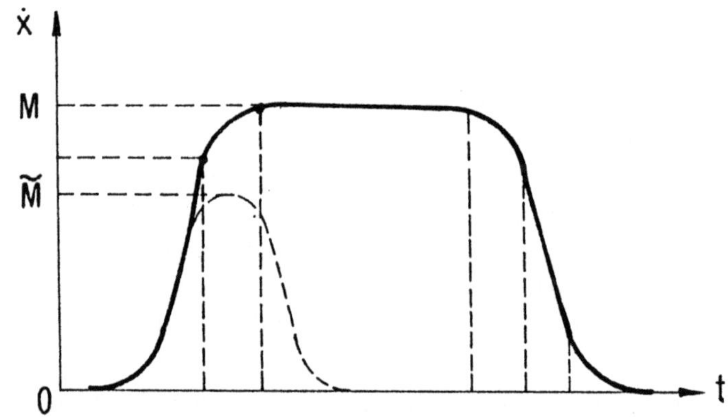

Figure 8.2. Deformed input control signal for position-type machine. $\tilde{M}$ is the maximum value of the human velocity for the current movement.

where x and y are the input and output signals of machines, respectively (see Chapter 4), and N^* is the limit value of a control handle shift. A targeting, one-coordinate task with an immobile target was considered.

8.2.2. POSITION SYSTEM MODELING

A time-optimal control law (Pontryagin *et al.*, 1965) for a position system (8.2) appears in Figure 8.1. Its deformed shape, which is due to HO limitations (8.1), is shown in Figure 8.2.

Based on a shift distance D, an output value y, and a gain k, this control law may be presented as one of three possible kinds of trajectories.

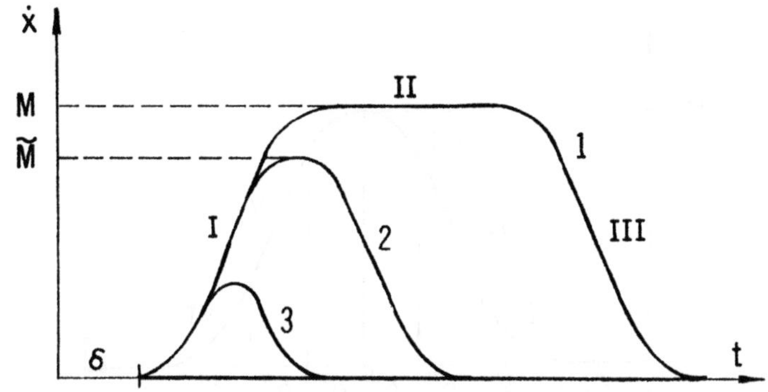

Figure 8.3. Three kinds of input control signals for a position-type machine. For curve 1, I is an acceleration part, II is a part with constant velocity, and III is a deceleration part.

The most complete one is shown as curve 1 in Figure 8.3.

While increasing the gain value from 0 to

$$k = k_1 = \frac{DLR}{M(R^2 + LM)}, \quad D = const,$$

in part II, the duration value gradually decreases, and, when

$$k = k_2 = \frac{DL^2}{2R^3},$$

the control law is described by curve 2 (Fig. 8.3). If we further increase $k (k > k_2)$, control is realized, as represented by curve 3 (Fig. 8.3), which is characterized by an absence of parts with constant values of control-signal velocity and acceleration (deceleration). It seems to us that there is an immediate connection between the bell-shaped trajectory of curve 3 and the trajectory obtained by means of the minimum-jerk principle (Flash and Hogan, 1985).

Thus, by knowing the HO control law, we may calculate a corresponding machine output signal based on machine equation (8.2), fitting the nearest parts of the control law to the corresponding boundary conditions. An example of a typical input signal curve 2 in Figure 8.3 is presented in more detail in Figure 8.4. A corresponding estimate of the maximum value of the task-fulfillment time appears as:

$$T = \delta_1 + \delta_2 + \frac{2R^2 + 2ML}{LR}. \tag{8.4}$$

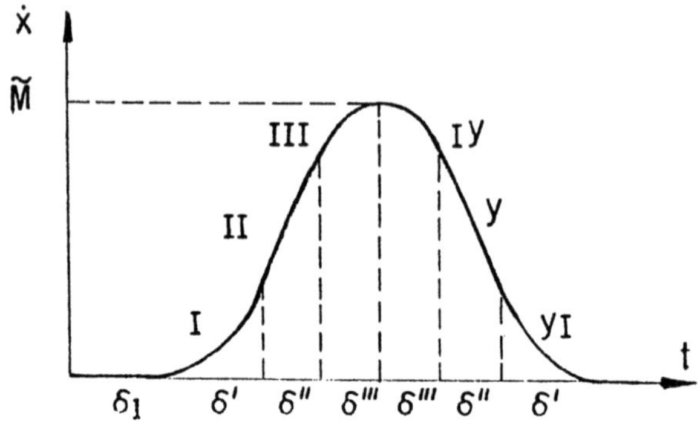

Figure 8.4. Different sections of position machine input control signal.

The full shift of the machine endpoint during this time is equal to D. This leads to the expression

$$D = k\tilde{M}\left(\frac{\tilde{M}}{R} + \frac{R}{L}\right). \tag{8.5}$$

This formula allows us to determine a condition for passage from one control law (Fig. 8.3) to another. Thus, the trajectory of curve 3 ensures a minimum shift of the machine endpoint. When endpoint shift for machine (8.2) is small,

$$D \leq kM\left(\frac{M}{R} + \frac{R}{L}\right),$$

the control law (Fig. 8.4) applies. When the shifts are large

$$D > kM\left(\frac{M}{R} + \frac{R}{L}\right),$$

the complete shape of the control law is as presented in Figure 8.2.

The equation for determination of value $\tilde{M}$ (Fig. 8.4) may be obtained from (8.5):

$$\tilde{M}^2 + \frac{\tilde{M}R^2}{L} - \frac{DR}{k} = 0.$$

A positive root of this equation,

$$\tilde{M} = \frac{-R^2 + \sqrt{R^4 + 4L^2DRk^{-1}}}{2L},$$

may be used to obtain a final form of the task-completion time expression (8.4)

$$T = 2RL^{-1} + \frac{\sqrt{R^4 + 4L^2DRk^{-1}} - R^2}{RL} + \delta_1 + \delta_2. \tag{8.6}$$

According to our approach, the task error is due to HO inexactitude when performing changes at a final stage of the control law:

$$\Phi = k\delta_2 \tilde{M} = k\delta_2 \frac{\sqrt{R^4 + 4L^2DRk^{-1}} - R^2}{2L}. \tag{8.7}$$

By applying this approach to other kinds of control laws (Fig. 8.3), we obtain for the range

$$0 \leq k < k_1 = \frac{DLR}{M(R^2 + LM}$$

the following expressions of criteria estimates:

$$T = \delta_1 + \delta_2 + D(kM)^{-1} + RL^{-1} + MR^{-1},\ \Phi = k\delta_2 M; \tag{8.8}$$

when

$$\frac{DLR}{M(R^2 + LM)} \leq k < \frac{DL^2}{2R^3}:$$

$$T = \delta_1 + \delta_2 + \frac{2R}{L} + \frac{\sqrt{R^4 + 4L^2DRk^{-1}}}{LR},\ \Phi = k\delta_2 \left(\sqrt{\frac{R^4}{4L^2} + RDk^{-1}} - \frac{R^2}{2L}\right); \tag{8.9}$$

when

$$\frac{DL^2}{2R^3} \leq k < \infty:$$

$$T = \delta_1 + \delta_2 + 4\sqrt[3]{\frac{D}{2kL}},\ \Phi = 2k\delta_2 \sqrt[3]{\frac{D^2L}{4k^2}}. \tag{8.10}$$

By just excluding parameter k from formulas (8.9) and (8.10), the trade-off error-time hyperbola may be obtained in a manner similar to that shown in Figure 1.3.

Formulas (8.9) and (8.10) may be used to built resultant dependences of the error and task fulfillment time for the machine (8.2). They are presented as functions of the machine gain in Figure 8.5. Note that they have a similar shape to those experimentally obtained for a position-type system presented in Figure 3.3. These dependences may be used to determine, for instance, an optimal value for the machine gain.

8.2.3. RATE SYSTEM MODELING

For a rate-type machine (equation (8.3)), an optimal shape for the control signal would resemble the curve shown in Figure 8.6. This follows from

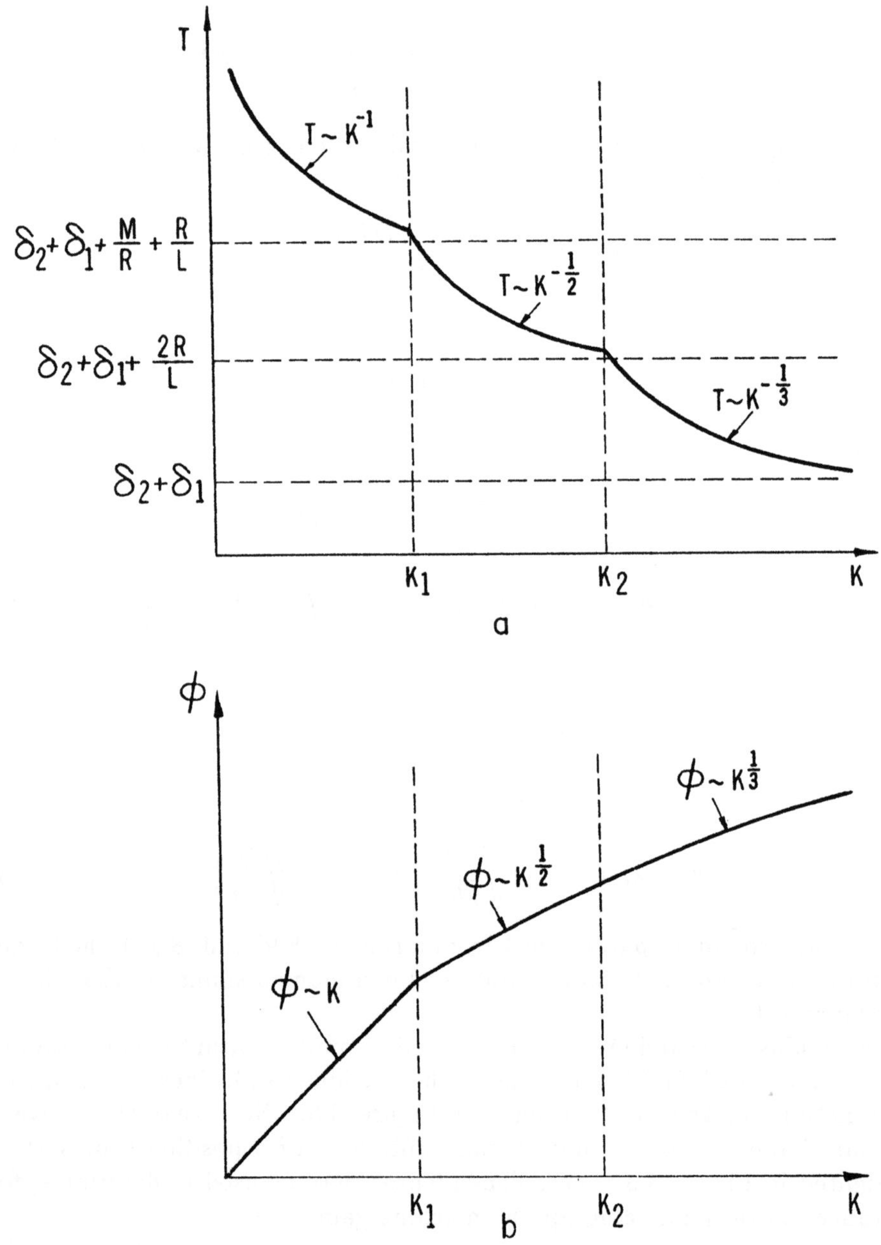

Figure 8.5. Quality criteria dependences (the task completion time T and error Φ) from position machine gain k.

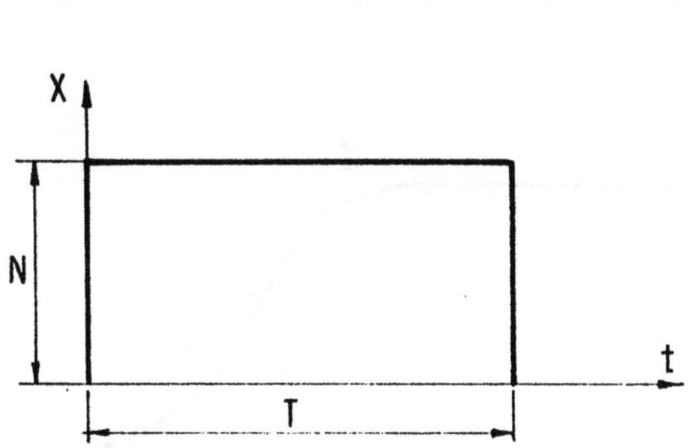

Figure 8.6. Optimal control input signal for rate-type machine (N is the maximum value of a control handle shift).

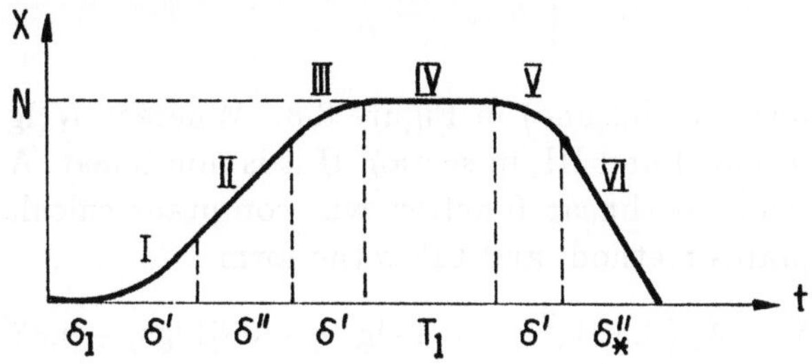

Figure 8.7. Deformed input signal for rate-type machine.

both experimental trials and from corresponding optimal problem solving (Pontryagin *et al.*, 1965). Time T of the optimal process is equal here to $D(\eta N)^{-1}$. As a consequence of HO limitations (8.1), this control trajectory is deformed and takes the form shown in Figure 8.7.

According to the methods described above, a value for T_1 may be obtained, following the appropriate calculations:

$$T_1 = \frac{D}{\eta N^*} + \frac{M^3}{24R^3N^*} - \frac{M}{R} - \frac{N^*}{M}.$$

The task-fulfillment time (Fig. 8.7) is

$$T = \delta_1 + \delta_2 + 3\delta' + \delta'' + T_1 + \delta''_* \qquad (8.11)$$

and the positioning error is

$$\Phi = \eta N^* \delta_2. \qquad (8.12)$$

Concerning equations (8.11) and (8.12), it is necessary to account for the experimentally revealed dependence of the maximum value N of control signal x from machine gain η. This statistically derived dependence is

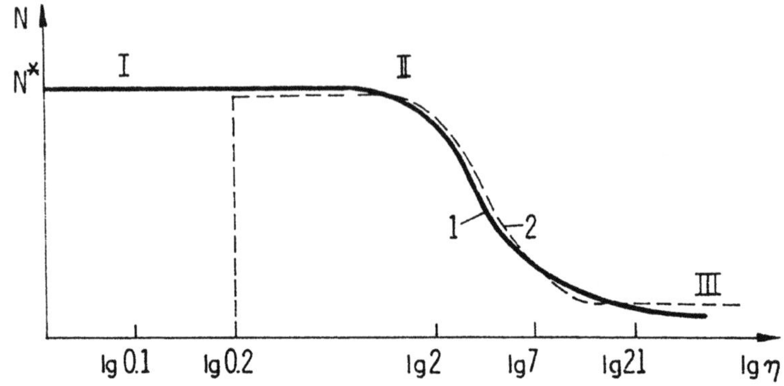

Figure 8.8. Dependence of the maximum value N^* of the control signal from the rate machine gain η.

presented by line 1 (solid line) in Figure 8.8. Whereas $N(\lg\eta)$ is constant in nature in sections I and III, in section II it is non-linear. A polynom that approximates this non-linear function was computer calculated by means of the least squares method, and takes the form

$$N = N^*(0.33\lg^3\eta - 0.44\lg^2\eta - 0.61\lg\eta + 0.88), \tag{8.13}$$

where N^* is the maximum magnitude of control signal x. Curve 2, corresponding to this equation, is shown in Figure 8.8 (dotted line).

On this basis, it may be concluded that the HO controls the inertialess machine with rate control as follows:

$$N = \begin{cases} N^* & \text{if } 0 < \eta \leq 0.2, \\ N^*(0.33\lg^3\eta - 0.44\lg^2\eta - 0.61\lg\eta + 0.88) & \text{if } 0.2 < \eta \leq 32, \\ N_\Delta & \text{if } \eta > 32. \end{cases}$$

Here N_Δ is again the resolution of the system input signal, which may be treated similarly to the way it was treated in Subsection 4.4.2.

By substituting the values in formulas (8.11) and (8.12) with those above, we can obtain final expressions of the quality criteria, the computer calculated graphs of which are presented in Figure 8.9. These graphs are successfully fitted to experimentally obtained corresponding dependences (Fig. 3.3).

Note the similar nature of these dependences, revealed for different types of performed remote control operations: reference trajectory tracing (Kobrinskii *et al.*, 1971), and targeting tasks (Kobrinskii *et al.*, 1974). In the latter work, it was explained that a trajectory tracing operation may be considered to be sequentially performed targeting tasks carried out on different points of the reference trajectory. So, the results obtained are rather

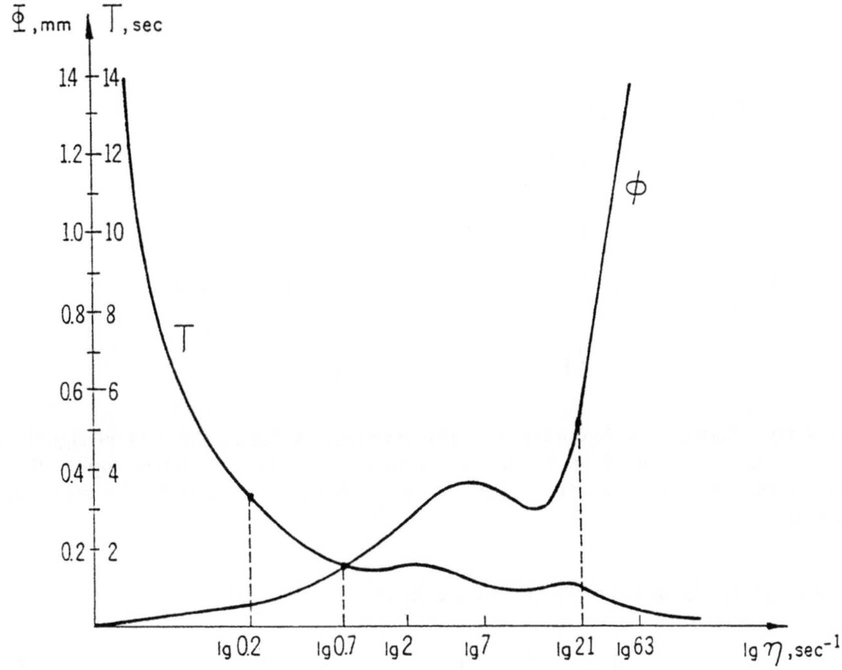

Figure 8.9. Dependences of quality criteria (the error Φ and task completion time T) from the rate machine gain η based on the experimentally obtained function $N(\lg \eta)$.

general in nature and may also be used, for example, to determine machine optimal parameters.

At the same time, a shortcoming of the approach presented here is its use of the approximation model (8.13), which does not reveal the nature of the human-control process. Therefore, we present yet another method for determining the dependence (8.13), one that takes into account the physical basis of human control processes.

This method is grounded in our suggestion that the HO chooses his/her motion amplitude N (Fig. 8.7) for the rate control by solving certain optimization problems. We assume that his/her objective function is the same one we considered previously for different stages of our work; that is, the complex criterion (3.4). By this we mean that, according to our suggestion and based on our described experimental results, the HO performs a certain operation by minimizing simultaneously both positioning error and task fulfillment time.

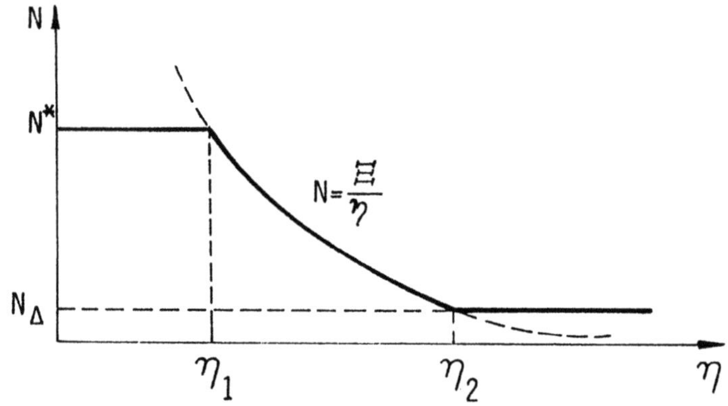

Figure 8.10. Calculated dependence of the maximum value of the control signal from the rate machine gain η. Here N^* is the maximum limit of the control signal; that is, the maximum shift of the control handle; and N_Δ is equal in value to the input signal resolution.

If so, from the above methods we have

$$\Phi = \delta_2 N\eta, \ T = \frac{D}{N\eta} + \delta_1 + \delta_2. \tag{8.14}$$

We substitute these expressions into the complex criterion

$$Q = \lambda_1 \Phi + \lambda_2 T, \ \lambda_1 + \lambda_2 = 1$$

and minimize it in the parameter N. This parameter optimal value is then

$$N = \Xi\eta^{-1}, \ \Xi = \sqrt{\frac{\lambda_2 D}{\lambda_1 \delta_2}}. \tag{8.15}$$

Thus, the dependence of the control signal maximum value N from the gain η may be graphically expressed as shown in Figure 8.10.

Once the value for N (8.15) is determined, the obtained control law (Fig. 8.11) is considered again. While calculating quality characteristics, we are taking into account that errors in switching control signals occur here twice. This dependence is now used in considering the expression (8.15). Thus, the final equations for the desired quality indices are:

for $N = N^* = const$

$$\Phi = \delta_2 N^* \eta, \tag{8.16}$$

$$T = \frac{D}{N^*\eta} + \frac{M^3}{24R^2N^*} + \frac{N^*}{M} + \frac{M}{2R} + \delta_1 + 2\delta_2;$$

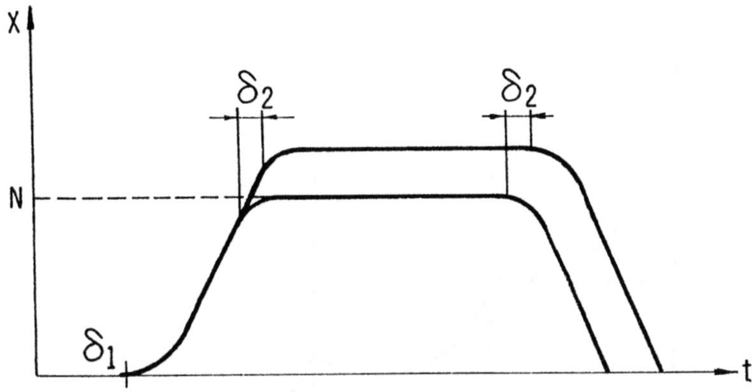

Figure 8.11. Scheme to calculate error and time of rate machine control process (δ_1 is a pure delay, δ_2 are errors in time of control signal switchings.

for $N = \Xi\eta^{-1}$

$$\Phi = (\frac{D}{\Xi} - \frac{M}{R})M\delta_2\eta + \Xi\delta_2 + \frac{M^4\delta_2\eta}{24R^2\Xi}, \quad (8.17)$$

$$T = \frac{M^3\eta}{24R^2\Xi} + \frac{\Xi}{M\eta} + \frac{M}{2R} + \frac{D}{\Xi} + \delta_1 + 2\delta_2;$$

for $N = N_\Delta = const$

$$\Phi = \delta_2 N_\Delta \eta, \quad (8.18)$$

$$T = \frac{D}{N_\Delta\eta} + \frac{M^3}{24R^2N_\Delta} + \frac{N_\Delta}{M} + \frac{M}{2R} + \delta_1 + 2\delta_2;$$

Graphs $\Phi(\eta)$ and $T(\eta)$ are shown in Figure 8.12 based on the above formulas.

Inspection of these graphs indicates that they are a quantitative fit with those of Figure 8.9 and with the experimental dependences (Fig. 3.3). Therefore, this approach may be further developed to solve the MMS synthesis problem, particularly those of telerobotic systems.

8.3. Method Based on the Study of Human Control Movement Structure

Results described in this Section have been obtained at Ben-Gurion University of the Negev during a joint investigation by Irina Gurevich, Yael Edan, and myself (Slutski *et al.*, 1997).

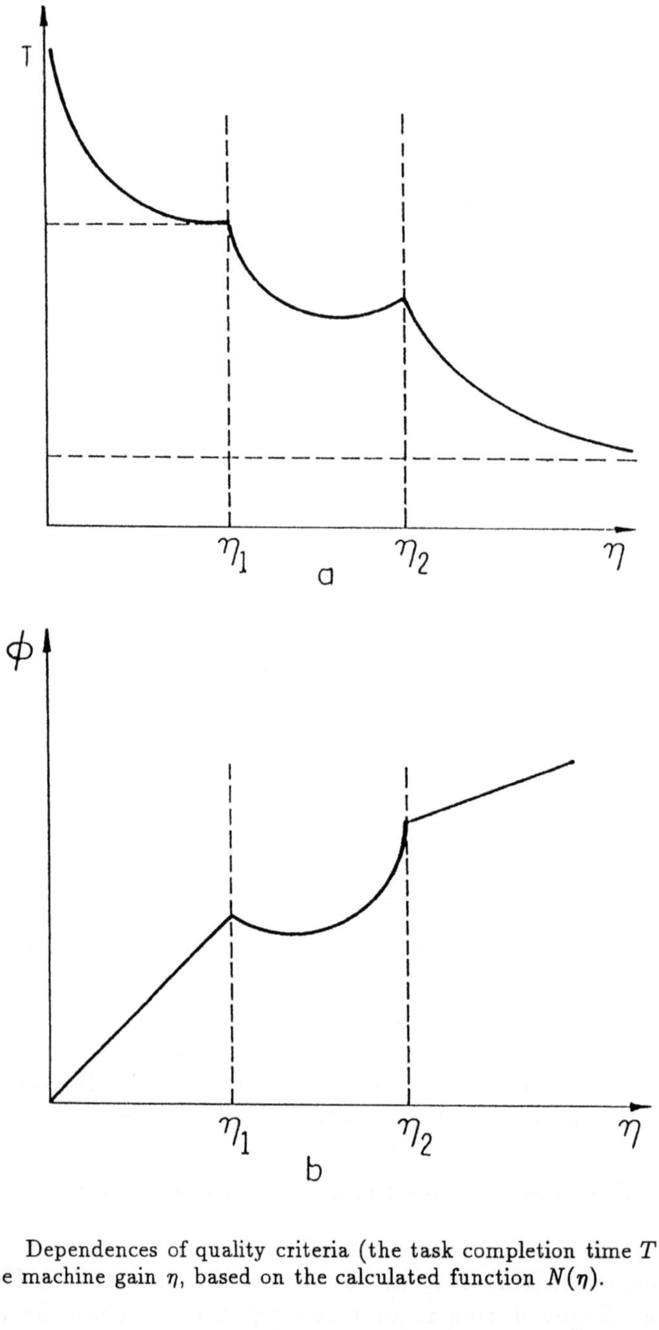

Figure 8.12. Dependences of quality criteria (the task completion time T and error Φ) from the rate machine gain η, based on the calculated function $N(\eta)$.

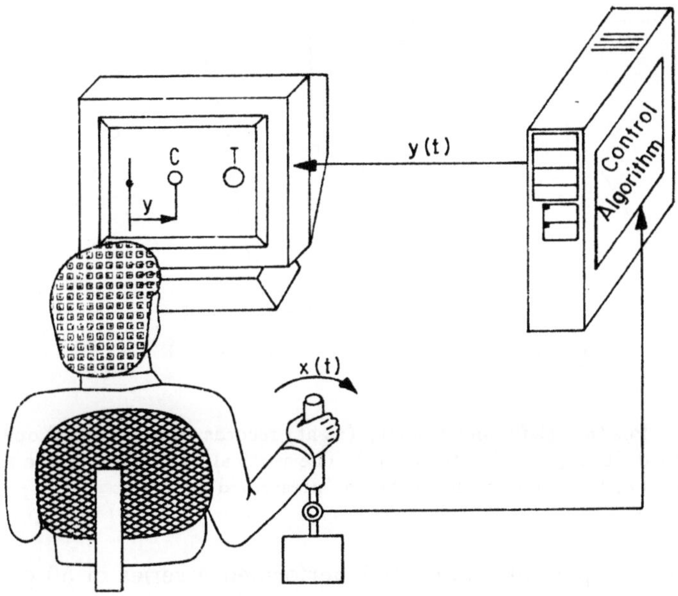

Figure 8.13. Experimental unit layout for human remote control investigation. Designations: C is a cursor, T is a target, $x(t)$ is a joystick output signal, and $y(t)$ is a cursor coordinate on the screen.

8.3.1. METHODS

Experiment Organization

This study investigated a class of telerobotic adaptive control systems in comparison to a conventional coordinate one. Concretely, the algorithm (6.17) with an automatic adjustment of parameter signal (Section 6.2.5) has been studied (when $k_{min} = 0$ for simplicity) comparing with the conventional position coordinate control algorithm (6.1). For the current research, it is vital that algorithm (6.17) presents an immediate transformation of one signal to another that permits the application of traditional approaches to analyze the control signal structure.

Six adult subjects (males), 28-49 years of age, participated in this study. Each HO had to control the shift of a cursor on one axis of the computer display by means of a joystick (Fig. 8.13).

The HO's task was to move cursor C from an initial position to a target T (Fig. 3.17) like in other investigations. The task parameters were chosen constant in our study; that is, the cursor size was 2.5 mm, the target size was 6.0 mm, and the cursor shift was 240 mm.

The HO was instructed to move the cursor to the target as fast and

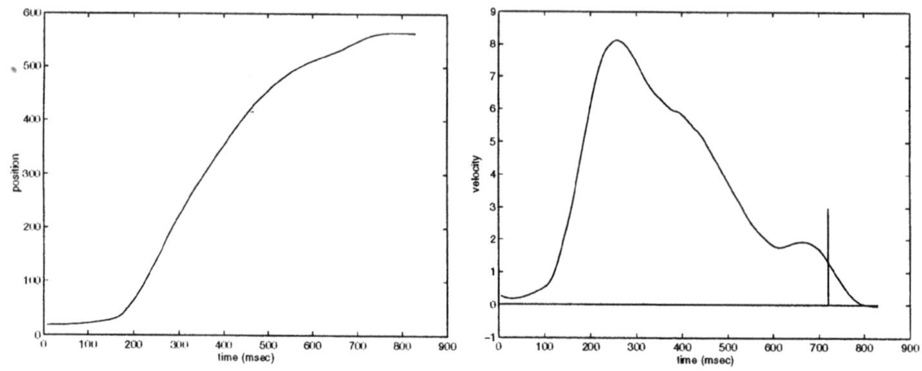

Figure 8.14. Position (left) and velocity (right) records from a typical operator movement performed during coordinate control. The vertical line indicates the instant when the cursor reaches the target region with the prescribed accuracy.

as accurately as possible. Each HO performed a series of 50 guidances to the target (a block of trials) for each type of control system. Each block of trials was recorded after one training session consisting of 50 guidances.

Both algorithms (6.1) and (6.17) were implemented on a PC 486 connected to a joystick with potentiometer transducers. The joystick had 4 controlled DOF and was connected to the PC via an analog to digital data acquisition card. It should be noted, by the way, that a stability problem of the adaptive systems was experimentally solved in this investigation by choosing a corresponding limit of the parameter α (6.17) to remove an instability danger of the system.

Data Processing

The joystick coordinates were sampled at 200 Hz using dedicated software written in Borland C++. They were further lowerpassed and filtered using the Butterworth digital filter. The joystick velocities were derived and smoothed. All measured movements that did not reach the end target with prescribed accuracy (5% of the total distance of the cursor shift) were discarded from the analysis. We did this with the intention that only skilled movements would be analyzed, and subjects who failed this criterion were not regarded as being skilled.

A characteristic view of the signal, realized by the HO when he is in control of the system with the implemented algorithm (6.1), is shown in Figure 8.14. This signal shape is similar to that obtained in (Keele, 1968; Kobrinskii *et al.*, 1974; Milner and Ijaz, 1990). Therefore, for future analysis, it was assumed that the process involves a sequence of elementary submovements with symmetrical velocity profiles, while the front of the first submovement forms the front of the whole control process.

The data processing involved the following stages:

1. The time of the control process initialization, its first peak, and the process end were distinguished. The first intersection of the transfer process graph with target line was taken as the moment of the process end. Based on the first two values, the first submovement was cut from the whole process.
2. The first step was subtracted from the control process to obtain the remaining part.
3. Items 1,2 were applied to the remaining part until a corresponding shift of the output point for a certain submovement became less than a selected accuracy.

As a result, the entire motion of the HO is represented as a sequence of partially superimposed submovements (Fig. 8.15).

That the control process trajectories are similar was established when the corresponding experimental data for the adaptive algorithm (6.17) were analyzed. A typical view of the signal, realized by one HO while in control of the system with algorithm (6.17), is shown in Figure 8.16. The signal is characterized by sharper oscillations of control signals as compared with the control signal of the position coordinate system (Fig. 8.14). These methods of velocity profile processing were also applied to adaptive system processes.

The outlined method of processing determined a number of structural characteristics of human control processes described in the next section.

Structural Characteristics of HO Control Process

The following characteristics were computed for each submovement for all recorded movements that reached the target with the prescribed accuracy:

- A distance proportion

$$p_{d_i} = \frac{D_i}{\Delta_i}, , \tag{8.19}$$

  where D_i is the distance that travels the i-th submovement, $\Delta_i = D - \sum_{j=1}^{i-1} D_j$ is the remaining distance to the target (similar to (Keele, 1968; Meyer *et al.*, 1988)), and D is the distance from the start to the target (see Fig. 8.15);
- A relative time delay

$$r_{t_i} = \frac{dT_i}{T_{i-1}}, , \tag{8.20}$$

  where dT_i characterizes the time interval between the beginning of the $(i-1)$-th and i-th submovements, and T_{i-1} is the duration of the $(i-1)$-th submovement;

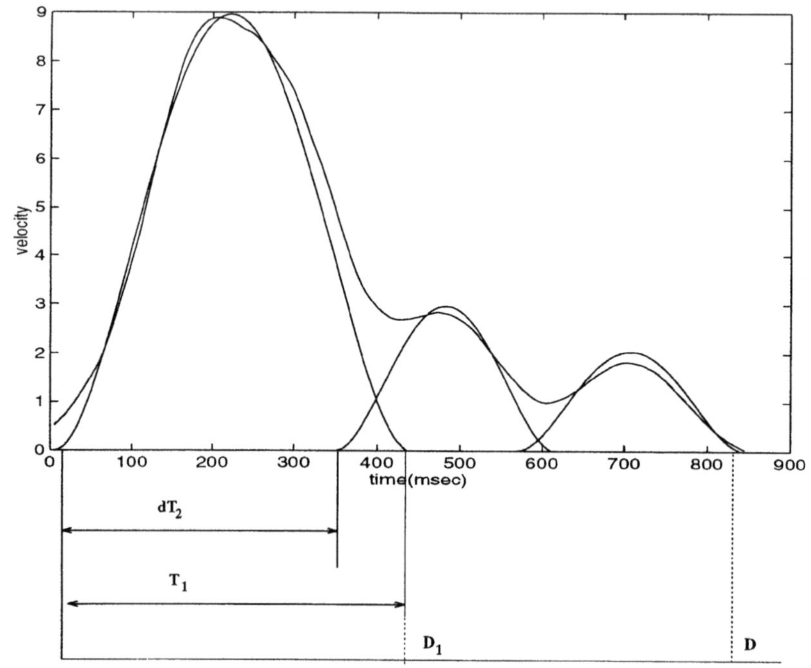

Figure 8.15. Velocity profile of a typical HO movement with the derived submovements. Here D corresponds to the total distance to the target, T_1 is the first submovement duration when the traveled distance is D_1, and dT_2 represents the time shift between the beginnings of the first and second submovements. ©1997 IEEE. Reprinted, with permission from (Slutski *et al.*, 1997).

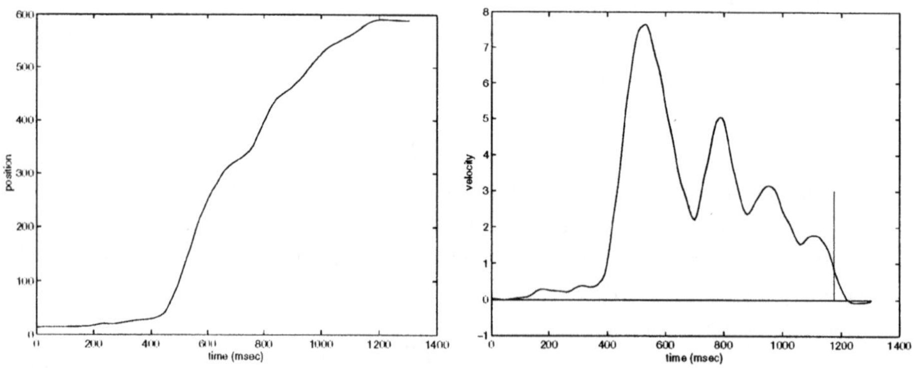

Figure 8.16. Position (left) and velocity (right) records from a typical operator movement performed during adaptive control. The vertical line indicates the instant when the cursor reached the target region with the prescribed accuracy. Right picture - ©1997 IEEE. Reprinted, with permission from (Slutski *et al.*, 1997).

- A maximum acceleration a_{max_i} of the i-th submovement;
- The number of submovements ns constituting each individual movement from the start to the target position.

Calculations of the above characteristics were used to fix some structural peculiarities of human control motions.

8.3.2. ANALYSIS OF EXPERIMENTAL RESULTS

In 58.6% of all recorded trials, the movements have reached the target within the prescribed accuracy and were considered sufficiently skilled.

For both the coordinate and adaptive systems, the one-step processes form only 7% of the total number of motions, and thus it seems that one-step processes are perhaps accidently realized.

The mean values of the described characteristics for the first submovements of each block of trials are presented in Table 8.1.

TABLE 8.1. The average values of HO motion characteristics for coordinate and adaptive control

Subjects	Control method	Characteristics		
		p_d	r_t	a_{max}
1	Coordinate	0.665	0.657	0.478
	Adaptive	0.592	0.671	0.494
2	Coordinate	0.688	0.65	0.426
	Adaptive	0.646	0.651	0.787
3	Coordinate	0.682	0.692	1.35
	Adaptive	0.586	0.679	0.671
4	Coordinate	0.633	0.64	0.837
	Adaptive	0.498	0.603	0.602
5	Coordinate	0.697	0.701	0.935
	Adaptive	0.645	0.675	0.472
6	Coordinate	0.619	0.624	0.812
	Adaptive	0.574	0.805	1.072

The differences between the obtained characteristics corresponding to different tested control algorithms were examined. Statistical analyses were applied to the values of the following characteristics: submovement number ns, and the parameters p_d, r_t, and a_{max}, corresponding to the first submovement within each individual guidance operation, obtained from 12 blocks of trials performed by six subjects.

Duncan's multiple-range test applied to the above characteristics grouped all six subjects together with respect to the mean value of the characteristics for each control algorithm tested. Statistical results obtained from Duncan's test are presented in Table 8.2.

TABLE 8.2. Duncan's test results of the movement characteristics (the number N corresponds to the numbers of blocks of trials performed by all the subjects, $\alpha = 0.05$)

Movement characteristics	Duncan grouping	Mean	N	Control algorithm
p_d	A	0.666	86	Coordinate
	B	0.589		Adaptive
a_{max}	A	0.770	86	Coordinate
	B	0.646		Adaptive
r_t	A	0.687	79	Coordinate
	A	0.657		Adaptive
ns	A	4.808	86	Coordinate
	A	4.453		Adaptive

Analysis results indicate that, at least for the first submovements, the characteristics p_d and a_{max} corresponding to the coordinate control algorithm (6.1) were significantly different from those obtained for the adaptive control algorithm (6.17). On the other hand, indices r_t and ns did not change significantly for the two tested control algorithms.

In summary, it was shown that, for the adaptive control algorithm,

1) index p_d decreased significantly relative to the coordinate system (11.56% on average for all tested subjects);

2) the average value of maximum acceleration among all tested subjects was 16.1% less than those obtained for the coordinate system.

Thus, the above data indicate that the values of indices p_d and a_{max} are primarily changed by the HO when switching to a new control mode.

8.3.3. DURATION ESTIMATE OF REMOTE CONTROL PROCESS

Estimate Derivation

Our experimental results have pointed to certain regularities observed in HO control motions. More specifically, the obtained results indicated that each operator movement can be described as a sequence of submovements, where each submovement travels a nearly constant proportion of the remaining distance to the target, while the time intervals between the beginnings of consequent submovements (i-th and $(i+1)$-th) (dT_i) can be approximately expressed as a constant proportion of the i-th submovement duration T_i.

Based on these results, the expression for the control operation performance time as a function of task parameters, such as a total distance to the end target and the above-described movement characteristics, can be obtained (Slutski *et al.*, 1997).

The total duration of the control process described as a consequence of n-submovements can be expressed (not considering a reaction delay time (Book, 1985)) as:

$$T(n) = dT_1 + dT_2 + ... + dT_{n-1} + T_n \ , \tag{8.21}$$

where T_n is a duration of the last, n-th submovement.

Substituting (8.20) in (8.21), we obtain:

$$T(n) = r_t \sum_{i=1}^{n-1} T_i + T_n. \tag{8.22}$$

On the other hand, we have shown (Gurevich *et al.*, 1995) that each submovement can be successfully accounted for by the minimum-jerk model (Flash and Hogan, 1985); that is, it may be described as follows:

$$x(t) = x_0 + (x_0 - x_f)(15\tau^4 - 6\tau^5 - 10\tau^3), \tag{8.23}$$

where $\tau = \frac{t}{t_f}$, $x(t)$ is the current position of the HC, x_0 is the initial position coordinate at $t = 0$, and x_f is the final position coordinates at $t = t_f$. It has also been shown (Flash and Hogan, 1985) that the duration of the i-th submovement is expressed as

$$T_i = \sqrt{\frac{5.7735 D_i}{a_{max}}}. \tag{8.24}$$

This expression establishes a connection between temporal and spatial characteristics of the control process.

The distance traveled by the i-th submovement was computed according to (8.19) as

$$D_i = p_d(D - \sum_{j=1}^{i-1} D_j). \tag{8.25}$$

Substituting (8.24) and (8.25) into (8.22), the duration of the n-step control process can be rewrited as:

$$T(n) = A\sqrt{Dp_d}[r_t S + \sqrt{(1 - p_d)^{n-1}}], \tag{8.26}$$

where

$$A = \sqrt{\frac{5.7735}{a_{max}}}, \; S = \frac{1 - \sqrt{(1 - p_d)^{n-1}}}{1 - \sqrt{1 - p_d}}.$$

When $n \to \infty$, we obtain an asymptotic estimate of this index that is independent of the submovement number in the structure of the control process as follows:

$$T(\infty) = \frac{Ar_t\sqrt{Dp_d}}{1 - \sqrt{1 - p_d}}. \tag{8.27}$$

The accuracy of (8.27) was evaluated for the different submovement numbers as:

$$E(n) = \frac{T(\infty) - T(n)}{T(n)} \cdot 100\%.$$

The duration calculation errors for different numbers of submovements in the control process structure are presented in Table 8.3, The average

TABLE 8.3. Duration calculation errors according to formula (8.27) obtained for different numbers of submovements

Submovement number, n	3	4	5	6	7
Error, $E(n), \%$	10.8	5.58	2.93	1.6	0.85

number of steps performed by the HOs while precise targeting was, as already noted, $ns = 4.5$ (Table 8.2). Therefore, from Table 8.3 may be seen that the estimated error of the duration calculation obtained according to (8.27) is relatively small for the average submovement number derived from our experimental data.

Estimate Application

The above results indicate that estimate (8.27) describes the change of the resultant characteristic of duration when the HO controlled parameters are changed. Therefore (8.27) was applied to estimate the adaptive system (6.17) time characteristics by comparing them with those of the coordinate coordinate system (6.1). The obtained results are united in Table 8.4, where T_a and T_s represent the durations of control processes according to experimental data and calculations based on expression (8.27), respectively. The values ΔT_a and ΔT_s are, accordingly, the absolute values of differences between indices T_a and T_s for coordinate and adaptive systems (i.e., an operational time decrease for the latter system). The average value of the control process time decrease for the adaptive system is 8.76%. This value,

TABLE 8.4. Experimental and calculated HO motion time characteristics

Subjects	System type	Characteristics			
		T_a	$\Delta T_a, \%$	T_s	$\Delta T_s, \%$
1	Coordinate	704.5		636.75	
	Adaptive	558.5	20.72	586.3	7.9
2	Coordinate	614.0		605.05	
	Adaptive	545.35	11.18	453.65	25.05
3	Coordinate	516.0		497.3	
	Adaptive	413.5	8.9	451.5	9.21
4	Coordinate	660.0		553.6	
	Adaptive	567.5	11.04	525.2	5.13
5	Coordinate	477.35		470.75	
	Adaptive	531.9	-11.4	543.5	-15.4
6	Coordinate	572.0		483.85	
	Adaptive	503.0	12.06	450.05	6.99
Average			8.76		6.48

which is calculated based on (8.27), gives a 6.48% duration decrease. The difference between these two values is 2.28%, which suggests a sufficient accuracy of calculation results obtained according to (8.27).

8.3.4. TIME-ESTIMATION OF THE SYSTEM WITH A NEW CONTROL ALGORITHM

Estimation Procedure

The obtained expression for movement duration presents an estimate that depends on the proposed characteristics of HO motion. As was shown previously, a control algorithm significantly affects the introduced characteristics p_d and a_{max}. Therefore, we have suggested that these parameters be adjusted by the HO to perform the task when he (she) uses different control algorithms. The idea was that the HO adjusts the "size" of the first submovement by changing p_d and a_{max}. In other words, the HO adjusts these parameters in order to obtain the best task performance for any concrete control algorithm.

To test this suggestion, we developed an optimization procedure that searched the appropriate scale factors by which the known values of p_d and a_{max}, derived for any conventional control mode, should be multiplied when passing to a control mode with a new control algorithm.

The studied coordinate control algorithm, with known characteristics for the sequence of overlapped minimum-jerk submovements, was used as a basis for the procedure development. New values of parameters p_d and a_{max} for the adaptive control algorithm (6.17) were found. The control processes durations obtained from this procedure were compared with those of the recorded HO control movements.

Simulation Procedure

The following assumptions and simplifications have been made in modeling the HO motions:

- Each submovement is generated according to the minimum-jerk model;
- Each submovement within a specific movement travels a constant proportion of the remaining distance to the target ($p_d = const$);
- Each submovement within a specific movement begins at a specific moment relative to the precedent movement duration ($r_t = const$);
- All submovements constituting a specific movement are generated with the same maximum acceleration ($a_{max} = const$).

This model of human motor behavior has been used into the developed procedure consisted of the following steps:

1. The velocity profile recorded for the coordinate control algorithm was analyzed. For the first submovement distinguished, which was approximated by the minimum-jerk model, a duration T_1 and traveled distance D_1 were obtained, and the parameters r_t, $p_d = \frac{D_1}{D}$, as well as a_{max} (see (8.24)) were computed.

2. The obtained parameters p_d and a_{max} were multiplied by scaling factors, which resulted in different trajectories of possible HO movements corresponding to the adaptive control system. These factors were chosen so as to minimize the end-position error with respect to the end target. To solve this minimization problem, the nonlinear optimization procedure was used.

3. Once the value of the scaling factors obtained from the above minimization procedure were derived, the resulting movement obtained for the adaptive control algorithm was simulated and its duration was determined.

4. The comparison was performed of the initial HO movement in the base coordinate control system with the simulated movement corresponding to the adaptive system.

Estimation Results

The above procedure was applied to the averaged values of proposed parameters of those operators who participated in our investigation. Figure 8.18 displays the sample velocity profiles of the simulated adaptive and actual coordinate movements.

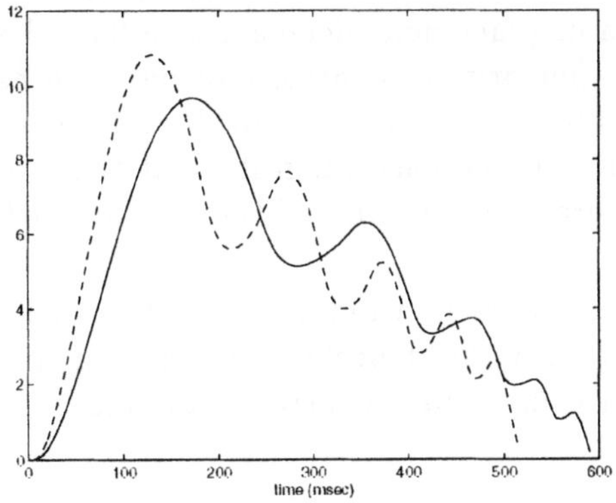

Figure 8.17. Velocity profiles of actual coordinate (solid lines) and simulated adaptive (dotted lines) movements for one operator. ©1997 IEEE. Reprinted, with permission from (Slutski *et al.*, 1997).

In Table 8.5 we present the simulation results of the adaptive control processes, together with the characteristics of the initial coordinate control processes and the actual adaptive control processes. It can be seen that the

TABLE 8.5. Values of the process duration T and handle displacement D for different control processes: C - actual coordinate control, A - actual adaptive control, S - simulated adaptive control

Subjects		1	2	5	6	Average
Indices						
	C	540	585	420	385	482.5
T	A	502	525	330	365	430.5
	S	475	525	365	355	430.0
	C	566	567	567	564	566.0
D	A	497	526	528	423	493.5
	S	554	550	563	546	553.25

simulation results revealed tendencies that hold for real control processes; namely, while using an adaptive method, the control process time decreased by 10.8% with respect to the experimental data, while the index obtained from simulation results was 10.9% less.

It is noteworthy that such a fast-operation improvement with a fixed accuracy level for the adaptive system was achieved under an amplitude decrease of HO arm motions. This index evaluation, according to the simulation data, shows a displacement decrease of 2.25%. Experimental results indicate that, when performing control motions, the handle displacement decreased by an average 12.8%. This result does not coincide very well with simulation data, but the decrease suggests that there may be some reduction of both the overall dimensions of the HC and HO operational zone dimensions.

In summary, the above method presents a possible way to estimate the movement duration without performing experiments by predicting HO motions in a system using a new control algorithm.

8.4. Discussion

The first approach presented here allows us to evaluate quality characteristics of remote systems with different kinds of machines, based on an adaptive model of human operator behavior. Its advantage is that, although it performs a rather approximate, preliminary estimate of these characteristics, it enables us to achieve principal analysis and synthesis of these systems, corresponding with remote system requirements of accuracy and fast-operation. This approach needs to be developed further and must be experimentally verified for machines with more diverse and complicated dynamics than those described in this study.

Another interesting direction of this approach is its potential development for application to analysis and design of adaptive systems with variable gain. Essentially this will require that some special alterations be made to the presented method.

The second methodology developed here is based on a highly precise and careful analysis of human motor behavior. It leads to a new, more precise, model of controlled HO motions in MMSs and, consequently, to a new estimate (8.27) of control process duration for these systems. This estimate, as well as Fitts's law, establishes a relationship between the control process duration and the distance to the target. However, this estimate has some features that may be explained by our experimental results.

Estimate (8.27) is based on the specific introduced indices that characterize the process structure. These characteristics include those known indices usually used to analyze HO control motions (p_d, number of submovements), as well as the indices following from the methods discussed above (such as r_t and a_{max}). It should be noted that the mathematical explanation of Fitts's law (Keele, 1968) assumes a constant time for each successful corrective submovement when each starts after the termination of a previous one. However, our results are unique in indicating the changing submovement duration and superimposed placement of submovements in the

HO control process structure. These findings differ from those obtained by Meyer et al. (1988) who reported discrete corrective submovements initiated successively, i.e., when the velocity of each previous movement segment has almost reached zero. Regarding the existence of superimposed submovements in an individual's arm movements, we refer to the results of Milner and Ijaz (1990). Thus, our contribution implies a quantitative estimation of the timing relationships within the extracted submovements. Such a layout provides a sufficiently good approximation of a real control process in time. The above relationships were then applied to the investigation of remote control processes.

Another subject of interest to this study was to account for the adaptive changes employed by human subjects to perform a new control task. To do this, the variations of the above indices due to application of different control algorithms were tested. It was found that the control algorithm does not affect the number of subsequent submovements in a body of an individual HO control motion, or their relative timing (parameter r_t). In contrast, the *"relative size"* of submovement determined by p_d and their *"absolute size"* determined by a_{max} were found to be affected by the control algorithm. The above results might suggest a possible adaptive strategy to be used by the HO in order to adjust his or her motor behavior to different control algorithms. According to the strategy suggested here, the relative submovement's size together with maximum acceleration might be treated as control parameters. This idea was supported by our simulations which permitted us, at least qualitatively, to account for the observed adaptive changes in HO control movements duration and movement characteristics.

Our results are consistent with earlier findings by Crossman and Goodeve (1983) who described adaptive changes in reaching wrist movements when the relationship between the visual and kinesthetic scales was altered. It can be observed from their results that, when the *"control/display ratio"* was altered, the relative sizes of the subsequent individual cycles change. For example, in the case of low gain values, they obtained one big cycle followed by several small cycles, while in the case of high gain values, the first cycle was relatively smaller, and the following cycles were larger. These results, although not treated in that way, are in good agreement with the adaptive strategy suggested here.

Contrary to Fitts's law, which describes certain relationships between movement duration and task parameters through the use of some positive constants that are determined by performing a multiple regression analysis, our estimate establishes the specific dependence of control process duration on a set of indices that account for the nature of HO control motions. This results in a new methodology for duration estimation that differs from that used in the variations of Fitts's law (Book, 1985; Book and Hannema, 1980), although there is an opportunity to develop the presented methodology and to use it similarly to methods applied in these previously performed investigations.

The described method of human movement analysis and the obtained estimate of the control process duration are of considerable practical interest and are very attractive for the evaluation of telerobotic system productivity. It has allowed us to establish that adaptive control of telerobot parameters is preferable to the conventional method of coordinate control, and also to establish a change tendency for parameters such as the HO operational zone dimensions.

Furthermore, a new design method for remote control systems has been proposed based on the new analysis of manual control processes. In line with the theory described above, a quality estimation of a newly developed telerobotic system is reduced to the prediction of new characteristics of HO control motion as compared to some existing system with known characteristics. Thus, final evaluation of the quality of a simulated process may be performed with the help of the asymptotic estimate (8.27). At this stage of development, the presented approach allows one to estimate a new process's tendencies at least qualitatively, whereas obtaining plausible quantitative estimates demands further development of the method. The proposed methodology may be expanded to include new elements that are contained in Fitts's law, such as target and cursor sizes. Such a methodology, after further development, would be capable of improving the design practice for telerobotic systems.

Especially promising is the possible combination of the first, more general, approach with the second methodology based on a detailed study of human motion structure. Such a combination may lead to the development of a universal method which will allow the solving of a wide range of different problems of MMS analysis and synthesis.

References

Book, W.J. (1985) Teleoperation arm design, Chapter 9. in R.C.Dorf and S.Y.Nof (eds.), *Handbook of Industrial Robotics. Intl. Encyclopedia of Robotics, Automation and Applications.* John Wiley & Sons, New York, pp. 138-157.

Book, W.J. and Hannema, D. (1980) Master-slave manipulator performance for various dynamic characteristics and positioning task parameters, *IEEE Trans. on System, Man and Cybernetics*, **Vol.SMC-10 no. 11**, pp. 764-771.

Crossman, E.R.F.W. and Goodeve, P.J. (1983) Feedback control of hand-movement and Fitt's law, *Quarterly J. of Experimental Psychology*, **35A**, pp. 251-278.

Djoldasbekov, U.A., Slutski, L.I., and Kalinich, S.S. (1986) Study of the quality of manual machine control, based on the human-operator optimal model, *Proc. of 8th Republican Inter-University Scientific Conf. on Mathematics and Mechanics*: Collection of scientific papers. Kazakh State University, Alma-Ata, pp. 8-13 (in Russian).

Elkind, J.J. (1964) A survey of the development of models for the human controller, in *Guidance and Control - 2.* New-York - London, pp. 623-643.

Fitts, P.M. (1954) The information capacity of human motor system in controlling the amplitude of movement, *J. of Experimental Psychology*, **Vol.47**, pp. 381-391.

Flash, T. and Hogan, N. (1985) The coordination of the arm movements: an experimentally confirmed mathematical model, *J. of Neuroscience*, **7**, pp. 1688-1703.

Gidikov, A.A., Marinov, M.P., and Guevski, I.V. (1973) Bionic aspects of time-optimal control problem, in *Problems of Bionics*. Nauka, Moscow, pp. 402-413 (in Russian).

Gurevich, I., Slutski, L., and Edan, Y. (1995) Analysis of remote control processes, *Proc. of 3rd French-Israeli Symp. on Robotics*, Herzeliya, Israel, pp. 97-101.

Keele, S.W. (1968) Movement control in skilled motor performance, *Psychological Bulletin*, **Vol.70 no.6**, pp. 387-403.

Kobrinskii, A.E., Sergeev, V.I., Slutski, L.I., and Tyves, L.I. (1971) On the quality evaluation of manual control systems, *Mashinovedenie* (Moscow), **no. 5**, pp. 3-10 (in Russian).

Kobrinskii, A.E., Sergeev, V.I., Slutski, L.I., and Tyves, L.I. (1974) Peculiarities of the operator fulfillment of the rough regulating motions in man-machine systems, *Mechanica Mashin* (Moscow), **Vol.46**, pp. 33-37 (in Russian).

Meyer, D.E., Abrams, R.A., Kornblum, S., Wright, C.E., and Smith, J.E.K. (1988) Optimality in human motor performance: Ideal control of rapid aimed movements, *Psychological Review*, **Vol.95 no. 3**, pp. 340-370.

Milner, T.E. and Ijaz, M.M. (1990) The effect of accuracy constraints on three-dimensional movement kinematics, *Neuroscience*, **Vol.35 no.2**, pp. 365-374.

Pontryagin, L.S., Boltyanskii, V.G., Gamkrelidze, R.V., and Mishchenko, E.F. (1965) *The Mathematical Theory of Optimal Processes*. Interscience Publishers, John Wiley & Sons, New York.

Slutski, L., Gurevich, I., and Edan, Y. (1997) Analysis and design of telerobot control based on human motor behavior study. *Proc. of 1997 IEEE Intern. Conf. on Robotics and Automation*, Albuquerque, USA, 20-25 April, pp. 1016-1021.

CHAPTER 9

CONCLUSION

A very large quantity of materials relating to the remote control of manipulators could not be completely illuminated here. We have described only the fundamentals of our subject due to the necessary limits on book lengths.

Thus, we have largely focused on a developed method of adaptive CPC for remote manipulators. The coordinate-parameter method represents a rather common approach to automatic machine control problem solving, but its use in telerobotic systems is now more advanced with respect to other applications of the method. This indicates that there are good prospects for the development of advanced teleoperation systems and other present-day machines, controlled by an HO. As examples of objects to which to apply this approach, we would like to distinguish such promising fields as Computer-Aided Teleoperation and Virtual Reality-Aided Teleoperation.

Maximum profit may be obtained while combining the proposed techniques for IR teaching with this process implementation in an environment of Virtual Reality. For instance, in recent years, simulation packages permit the effective simulation of robots and other equipment in 3D motion and provide rapid workcell layout and automatic program generation and verification.

Principally, about 80% of the work necessary to teach an IR can be done today off-line with the recent CAD systems; another 20% of the work may be performed by using either Virtual Reality environments, or on-line methods described in the book (according to data of P. Coiffet). However, it is important to note that the first 80% of the work and the work in Virtual Reality environments are performed with the help of remote control methods as well.

While using the off-line approach, the same kind of problem one has with an actual robot appears, only now has to point out a specific point on a computer display fast and with high accuracy. The same means of remote control are also used for such teaching of virtual IRs. In this case, the main advantages of this off-line approach, such as:

- a teaching process may be realized even without the existence of a real robot;
- the danger that equipment may be damaged during a teaching period is

avoided;
will be combined with the advantages of proposed on-line control technology, such as:
- the reduction of task completion time under given accuracy levels because of adaptive algorithm implementation;
- the use of more universal HCs.
The development of this approach may significantly change the state of the problem of IR teaching.

It seems to us that this approach also indicates a way to solve the problem of an optimal co-ordination of human and machine in a MMS. This problem has not yet been effectively solved, not for a lack of appropriated technical means, but rather, for the lack of a clear, formal determination of the problem itself.

In the context of the approach developed here, we introduce, in fact, a system of quality criteria and technical means in order to change values of these criteria for the systems. We may believe that when the quality criteria are optimized, an interface between the HO and machine will be optimal as well. Thus, we suggest that the solution obtained optimizes somewhat the MMS (e.g., HO - telerobot system) and, further, that this solution is flexible. It offers the possibility of quickly obtaining a specific technical solution; i.e., a new value of the system gain when system requirements are changed. These requirements may be expressed in a form of the complex criterion (3.3) or the set of weight coefficients of the criterion (3.4).

A number of problems have already been solved using the adaptive method, not only for the described tactical level of remote manipulation systems, but also for the lower, executing control level. Note that further prospects are related, first, to the practical use of the adaptive method for control of actual robots and manipulators of various levels of sophistication, and second, to the application of this method to the development of higher level robotic systems.

An interesting application of this approach might be developed in relation to a voice control. Thus, an ordinary voice control would be used as a coordinate control signal when the HO gives voice commands to a robot. The robot would then works at a nominal rate. At the same time, when a need arises to speed up the robot work rate, the HO would raise his/her voice, the voice level serving as a parameter signal that would change the robot work rate. This example of a work system now seems like an episode from science fiction, where, when a danger appears, the HO cries out a command to an intelligent robot and the robot darts off to perform this command.

No one should receive the impression that the control techniques presented in this volume are the only means proposed by the author. They are, in fact, only one of the elements of the basic control system that might be useful to explore in terms of overall improvement. Based on this element, further system improvement is possible by other means, which are traditionally applied in telerobotics (for instance, the force reflection, ergonomic design).

Regarding the method of adaptive semiautomatic control, it should be emphasized that it is, in fact, at an initial stage in its development. There are a number of things that have remained outside the frame of this book but that are important for method realization. Among these are:

- the choice of optimal means of implementation;
- exploration of the influence of object dynamics on adaptive control system characteristics;
- adaptive system stability,
- force control based on adaptive algorithms,
- development of combined adaptive systems.

This work is offered as a first step in this direction. The author hopes that this book will stimulate an interest in the development of these important and theoretically interesting problems and that other steps will follow in the future.

ABBREVIATIONS

CAD Computer-Aided Design

CPC Coordinate-Parameter Control

DC Direct Current

DOF Degree of Freedom

EE End Effector

EO Executing Organ

FM Flexible Master

HC Hand Controller

HO Human Operator

ID Input Device

IR Industrial Robot

MMS Man-Machine System

MSS Master-Slave System

SAC Semiautomatic Control

SOSF Sum of Squared Forces

2D Two-Dimensional

3D Three-Dimensional

6D Six-Dimensional

INDEX

International Series on
MICROPROCESSOR-BASED AND INTELLIGENT SYSTEMS ENGINEERING

Editor: Professor S. G. Tzafestas, *National Technical University, Athens, Greece*

1. S.G. Tzafestas (ed.): *Microprocessors in Signal Processing, Measurement and Control*. 1983 ISBN 90-277-1497-5
2. G. Conte and D. Del Corso (eds.): *Multi-Microprocessor Systems for Real-Time Applications*. 1985 ISBN 90-277-2054-1
3. C.J. Georgopoulos: *Interface Fundamentals in Microprocessor-Controlled Systems*. 1985 ISBN 90-277-2127-0
4. N.K. Sinha (ed.): *Microprocessor-Based Control Systems*. 1986 ISBN 90-277-2287-0
5. S.G. Tzafestas and J.K. Pal (eds.): *Real Time Microcomputer Control of Industrial Processes*. 1990 ISBN 0-7923-0779-8
6. S.G. Tzafestas (ed.): *Microprocessors in Robotic and Manufacturing Systems*. 1991 ISBN 0-7923-0780-1
7. N.K. Sinha and G.P. Rao (eds.): *Identification of Continuous-Time Systems*. Methodology and Computer Implementation. 1991 ISBN 0-7923-1336-4
8. G.A. Perdikaris: *Computer Controlled Systems*. Theory and Applications. 1991 ISBN 0-7923-1422-0
9. S.G. Tzafestas (ed.): *Engineering Systems with Intelligence*. Concepts, Tools and Applications. 1991 ISBN 0-7923-1500-6
10. S.G. Tzafestas (ed.): *Robotic Systems*. Advanced Techniques and Applications. 1992 ISBN 0-7923-1749-1
11. S.G. Tzafestas and A.N. Venetsanopoulos (eds.): *Fuzzy Reasoning in Information, Decision and Control Systems*. 1994 ISBN 0-7923-2643-1
12. A.D. Pouliezos and G.S. Stavrakakis: *Real Time Fault Monitoring of Industrial Processes*. 1994 ISBN 0-7923-2737-3
13. S.H. Kim: *Learning and Coordination*. Enhancing Agent Performance through Distributed Decision Making. 1994 ISBN 0-7923-3046-3
14. S.G. Tzafestas and H.B. Verbruggen (eds.): *Artificial Intelligence in Industrial Decision Making, Control and Automation*. 1995 ISBN 0-7923-3320-9
15. Y.-H. Song, A. Johns and R. Aggarwal: *Computational Intelligence Applications to Power Systems*. 1996 ISBM 0-7923-4075-2
16. S.G. Tzafestas (ed.): *Methods and Applications of Intelligent Control*. 1997 ISBN 0-7923-4624-6
17. L.I. Slutski: *Remote Manipulation Systems*. Quality Evaluation and Improvement. 1998 ISBN 0-7932-4822-2

KLUWER ACADEMIC PUBLISHERS – DORDRECHT / BOSTON / LONDON

International Series on

MICROPROCESSOR-BASED AND INTELLIGENT SYSTEMS ENGINEERING

VOLUME 17

The titles published in this series are listed at the end of this volume.

Remote Manipulation Systems

MIX
Papier aus verantwortungsvollen Quellen
Paper from responsible sources
FSC® C105338

If you have any concerns about our products,
you can contact us on
ProductSafety@springernature.com

In case Publisher is established outside the EU,
the EU authorized representative is:
Springer Nature Customer Service Center GmbH
Europaplatz 3, 69115 Heidelberg, Germany

Printed by Libri Plureos GmbH
in Hamburg, Germany